D1065188

DUEL AT DAWN

NEW HISTORIES OF
SCIENCE, TECHNOLOGY, AND MEDICINE

Series Editors

Margaret C. Jacob

Spencer R. Weart

DUEL AT DAWN

HEROES, MARTYRS, AND THE RISE OF MODERN MATHEMATICS

Amir Alexander

HARVARD UNIVERSITY PRESS

Cambridge, Massachusetts

London, England

2010

Printed in the United States of America

Library of Congress Cataloging-in-Publication Data

Alexander, Amir.
Duel at dawn : heroes, martyrs, and the rise of modern mathematics / Amir Alexander.
p. cm.
Includes bibliographical references and index.
ISBN 978-0-674-04661-0 (alk. paper)
1. Mathematics—Social aspects—History. I. Title.
QA10.7.A44 2010
510.9—dc22 2009043776

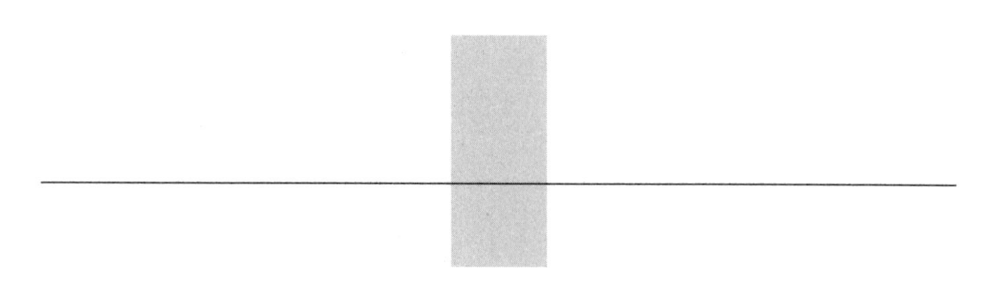

To Bonnie

"Beauty is truth, truth beauty,"—that is all
Ye know on earth, and all ye need to know.

—John Keats

CONTENTS

DUEL AT DAWN

INTRODUCTION: A SHOWDOWN IN PARIS

In the early dawn of May 30, 1832, in the southern Paris suburb of Gentilly, two young men faced each other with pistols drawn. The identity of one of the men has been lost to us, although he may have been Pescheux d'Herbinville, a radical revolutionary and member of a banned militia. The other man we know well: he was Évariste Galois, a political radical like his opponent, recently released from prison for political offenses. He was also, however, an aspiring mathematician possessed of uncommon gifts and startlingly original insights. According to a newspaper report, the two rivals had agreed beforehand that only one of the pistols would be loaded, leaving the outcome of the affair to pure chance. When the signal was given, a single shot shattered the morning stillness, and Galois fell to the ground, mortally wounded. He died the following day in a nearby hospital, in the arms of his younger brother Alfred. He was only 20 years old.

The duel that took place on that Parisian dawn is at the center of this book. Parts of what follows deal with events that took place long before that spring morning, going back to the seventeenth century. Other parts focus on the years surrounding the duel and those that followed, leading up to the present day. Along the way the reader will encounter some great mathematicians and a few small ones, national heroes and lonely outcasts, saintly innocents and scheming swindlers, princes, paupers, and martyrs. The story will range all across Europe and beyond—from the lights of Paris to the hinterlands of Transylvania, and from the frosts of Norway to the sunshine of Tuscany, with brief stops in Cambridge, St. Petersburg, Beijing, and Princeton, New Jersey. But the central argument of this book is simple and can be stated briefly: the duel that ended the life of young Galois marks the end of an era in the practice of mathematics and the beginning of another. In a word, it marks the birth of modern mathematics.

Associating a profound intellectual transformation in mathematics with a violent and altogether random event like a duel may seem surprising. After all, it is difficult to argue that if Galois had patched up his differences with d'Herbinville beforehand, or if he rather than his rival had been handed the loaded pistol, the entire course of the development of mathematics would have been different. That, indeed, seems unlikely. But the fact is that the tragic story of Évariste Galois, the mathematical genius who burned brightly but all too briefly, is not as unusual as one might think among the mathematicians of his and subsequent generations. It is, rather, the most famous and dramatic of an entire genre of mathematical stories that originated in the early decades of the nineteenth century but is still going strong today. It is a tale of rejection, disillusionment, and even martyrdom of the most creative mathematicians of the age. It is a tale, furthermore, that went hand in hand with the invention and development of a new type of mathematics.

Consider, for example, Galois' older contemporary, Niels Henrik Abel (1802–1829), who solved the long-standing mystery of the quintic equation but lived in penury and died of pulmonary tuberculosis at age 26. Or János Bolyai (1802–1860), a young Hungarian aristocrat who in his twenties developed the first non-Euclidean geometry, only to be driven from the field by the chilly reception accorded his discovery by established mathematicians. Some years later there was Srinivasa Ramanujan (1887–1920), a self-taught Indian genius who had an uncanny affinity for numbers and spent the years of the Great War in Cambridge. In England he contracted an undiagnosed illness, and he died within a year of returning to India in 1919. John Nash (born 1928), the subject of the best-selling biography and movie *A Beautiful Mind,* suffered from repeated bouts of paranoid schizophrenia before gaining his due recognition for his work in game theory late in his career.

Other examples include Kurt Gödel (1906–1978), "the greatest logician since Aristotle," who starved himself to death out of fear of poisoning; Alexander Grothendieck (born 1928), arguably the most influential mathematician of the mid-twentieth century, who disappeared from his home in 1991 and is reputedly living alone in a Pyrenees village; and Grigory Perelman (born 1966), conqueror of the Poincaré conjecture, a classic problem that had defied the best efforts of mathematicians for a century. Perelman turned down the Fields Medal (the mathematics equivalent of a Nobel Prize) and retired to his tiny apartment in St. Petersburg, claiming that rivals were trying to steal his credit. Considered in this distinguished company, the story of Galois seems not so much an unaccountable and extreme exception, but rather the most dramatic of a class of mathematical biographies, depicting the lives of some of the most distinguished mathematicians of the past two centuries. Among modern mathematicians, it seems, extreme eccentricity, mental illness, and even solitary death are not a matter of random misfortune. They are, rather, almost signs of distinction, reserved only for the most outstanding members of the field.

It may well be, of course, that the seeming prevalence of eccentricity and tragedy among modern mathematicians is no more than an illusion. The majority of practicing mathematicians are undoubtedly normal individuals, well-integrated into their family and community. Nevertheless, it is undeniable that in the popular imagination, as well as in the more famous mathematical biographies, mathematicians feature prominently as loners and misfits who never find their place in the world. This image is particularly striking when we consider that it is quite different from popular images of scientists in other fields, such as physics or biology. The celebrated physicists of the past century, for example, men like Albert Einstein, Niels Bohr, and Werner Heisenberg, may have been viewed as somewhat eccentric, but there is no touch of the tragic about them. To the contrary, they were all active men of affairs, prominent both socially and professionally.

The popular image of the tragic misfit not only sets modern mathematicians apart from practitioners of other sciences but also separates them from mathematicians of the past. As recently as the eighteenth century mathematicians were viewed in a very different light, as scientific, cultural, and even political leaders. Jean le Rond d'Alembert (1717–1783), for example, was not only one of the leading mathematicians of his era and a member of the Academy of Sciences but also the permanent secretary of the literary Académie française, a coeditor of the *Encyclopédie*, and the acknowledged leader of the French "philosophes." Leonhard Euler (1707–1783), perhaps the deepest and certainly the most prolific geometer of the Enlightenment, held the most prestigious mathematical chairs in Europe and was a personal acquaintance of the king of Prussia and the Russian empress. Joseph-Louis Lagrange (1736–1813) succeeded Euler in Berlin before moving to Paris, where during the French Revolution he became president of the committee on weights and measures and a founding member of the École polytechnique. Overall, it is fair to say that the narrative of the tragic mathematical misfit is completely absent from the biographies of the leading mathematicians of the Enlightenment.[1]

It is clear from all this that the iconic image of the tragic mathematician made its appearance quite suddenly, in the early decades of the nineteenth century. It centered on the figures of Galois, Abel, and Bolyai, and the legends that soon grew around them. The most famous of these, the story of Galois, spread slowly in mathematical circles in the decades after his death, but by the end of the century young Galois' reputation had undergone a stunning transformation: the lonely outcast, ignored by the mathematical establishment, had become an iconic figure of the field, a revered martyr to mathematics. As his posthumous fame grew, Galois became a model of the pure mathematical life, pursued for its own sake without regard for the consequences. Since that time each successive generation of mathematicians has produced its own famous examples of the type: a genius who had glimpsed the brilliance of the perfect world of mathematics, but is destroyed by life in the highly imperfect world of men.

The period at which the legend of the mathematical martyr made its appearance stands out in the history of mathematics for other reasons as well. The early decades of the nineteenth century mark the birth of modern mathematics as practiced to this day in academic departments. Only a generation earlier, Enlightenment geometers such as Euler, d'Alembert, and Lagrange practiced a mathematics that was tied, at its core, to the physical world. Even at its most abstract the work of these *grands géomètres,* as they were called, was ultimately derived from physical reality as it exists in the world around us.

But in the early decades of the nineteenth century a new generation of practitioners proposed a wholly new way of viewing and practicing mathematics. For Augustin-Louis Cauchy (1789–1857), Niels Henrik Abel, Évariste Galois, and a growing number of their colleagues, mathematics was not derived from the physical world but was, rather, a world unto itself. It was a wondrous alternative reality governed solely by the eternal laws of pure mathematics, unsullied by the crass realities of the world around us. Unlike their elders, the new mathematicians were not so much interested in acquiring new and useful results as they were focused on the internal architecture of mathematics itself, its interconnections, and the precise meaning of its statements. Mathematics, for them, was its own self-contained world and could be judged by mathematical standards alone.

This new approach did not take the mathematical world by storm, but it did make steady progress throughout the nineteenth century. By the turn of the twentieth century it set the standard for the entire international community of pure mathematics, and it remains to this day the defining view of academic mathematicians, instilled in incoming students in countless college-level and graduate courses. The approach is, in fact, so deeply ingrained in the culture of the field that it is hardly ever noticed or commented on. Most practicing mathematicians take it for granted that their subject matter is a Platonic world of pure mathematical objects, and that their mission is to uncover the hidden outlines of this mathematical reality. In this they are true heirs to the new mathematicians of the early nineteenth century, who transformed a field that was centered on the physical world and who established it on radically new foundations that have lasted to this day.

Remarkably, the new persona of the tragic mathematical misfit and the new practice of pure and insular mathematics came on the scene at precisely the same time. The central argument of this book is that this is no coincidence: the mathematical legend that appeared in the age of Galois is inseparable from the new mathematical practice that transformed the field in those years. The story and the practice emerged together in the early decades of the nineteenth century, and each supported the other. A radically new type of mathematics required a new type of mathematician, with a very different relationship to his craft, to the world, and to his fellow men. At the same time a new story of genius and martyrdom, drawn from the discourse of High Romanticism, legitimized and allowed for a new type of mathematical knowledge: impractical, self-referential, irrelevant to worldly life, and judged only by its purity, its truth, and its beauty. Since then the story of the mathematical martyr and the practice of the new self-contained mathematics have survived, and thrived, together.

The notion that mathematical practice is inseparable from popular stories may appear counterintuitive. After all, is not mathematics the most insular of sciences, the field most immune to the ephemeral influences of culture? Is it not, in effect, defined by the fact that it concerns itself solely with mathematical objects and follows only pure mathematical reasoning? To some extent these common objections reflect a particular view of mathematics that prevails at this particular historical moment. Indeed, the insistence on the absolute purity and insularity of mathematics is a feature of the way the field has been understood and practiced for the past two centuries. Earlier generations of geometers had different and often much broader understandings of the proper boundaries of mathematics. At most times they were far less insistent on the insularity of mathematics, and never before the nineteenth century were they as determined that mathematics should be judged exclusively by its own internal standards. In other words, the fact that we find it surprising that mathematics can be shaped by broader cultural concerns is, to an extent, a product of historical culture.

In fact, what is considered a proper "mathematical object" or a legitimate "mathematical argument," the boundaries of the field and its relationship to the outside world, can vary enormously over time. The mathematics produced by Archimedes in the third century BCE and that produced by Alexander Grothendieck in the twentieth century CE, for example, can both be recognized as mathematics, at least by the modern practitioner. They are both part of the same millennia-long tradition. Nevertheless, Archimedes would not recognize his subject in Grothendieck's practice, and Grothendieck would never consider doing mathematics in the manner of Archimedes. This is not simply because Grothendieck "knows more" than Archimedes, although he undoubtedly does. It is because the boundaries of the field and its method of argument—its subject matter, in short—have changed to a degree that would make it unrecognizable to an ancient geometer.

There is nothing within mathematics itself that can determine what the subject matter and boundaries of the field are to be. These fundamental features, which define the nature of mathematics, are an inseparable part of the historical moment in which they originate. Inevitably they are shaped by broader cultural trends, whether philosophical, literary, artistic, or even political, and mathematical stories like the legend of Galois are an indispensable guide in tracing them. Poised between technical mathematical practices, on the one hand, and broader cultural trends, on the other, mathematical stories have a stake in each. Like other literary creations, they are rooted in broader cultural trends, mirroring them while also contributing to them. At the same time the stories are also inseparable from the technical practice of mathematics and help shape a particular understanding and practice of the field. Mathematical stories, in other words, are a bridge between refined and highly abstract mathematical practices and the historical and cultural moment in which they are practiced. By following the evolution of these tales and their transformation over time, we are given access to a hidden history of mathematics, one in which even abstract technical developments are part of human culture and the broad sweep of history.

The modern legends associated with modern mathematicians such as Galois, Ramanujan, and Perelman and the recent success of movies such as *A Beautiful Mind* and *Good Will Hunting* are surely indications that the mathematical story is very much with us today. The origins of the genre, however, go back thousands of years, to classical Greece and likely earlier. In the fifth century BCE the Pythagorean philosopher Hippasus of Metapontum proved that the side of a square is incommensurable with its diagonal. This discovery was quickly recognized for its far-reaching implications, which thoroughly challenged the Pythagorean belief that everything in the world could be described by whole numbers and their ratios. Sadly for Hippasus, he did not live to enjoy the fame of his mathematical breakthrough. Not long after making his discovery, he traveled aboard ship and was lost at sea. Since that time different versions of the story have come down to us. In some versions Hippasus's "shipwreck" was contrived by his own Pythagorean brothers, who feared that his discovery would undermine their core beliefs, whereas in others he was not killed but only expelled from the brotherhood for his indiscretion in revealing its most profound secrets. But whatever version one adopts, it is clear that the story of Hippasus is not meant to be an accurate chronicle of a tragic event that took place 2,500 years ago, but a morality tale conveying deeply held truths about the meaning of mathematics and its potential dangers.

Other mathematical tales soon followed. Like Hippasus, the pre-Socratic philosopher Aristippus also suffered shipwreck, but his tale carries a very different moral. When he became stranded with his shipmates on an unfamiliar shore, he noticed geometrical diagrams drawn in the sand. Taking heart, Aristippus calmed his distressed companions and assured them that educated and humane men were nearby, and they would be saved. And so it proved to be.[2] Euclid, according to another popular tale, admonished King Ptolemy that "there is no royal road to mathematics," and Archimedes, a favorite hero of mathematical tales, ran naked through the streets of Syracuse shouting "Eureka!" when a sudden insight on buoyancy struck him during a sojourn in the public baths. Years later he was killed when, oblivious to the sack of his city, he asked an ignorant Roman soldier to wait while he worked out a geometrical problem.

There is more. In my previous book, *Geometrical Landscapes,* I showed how in the sixteenth and seventeenth centuries stories abounded about geometers as intrepid voyagers on the seas of mathematics. Inspired by the voyages of geographical exploration in which not a few of them played a role, mathematicians fashioned themselves as explorers in their own right, hazarding the shoals of paradox and seeking the safe harbor of geometrical demonstration. Such stories not only placed the practice of mathematics squarely at the heart of the heroic age of exploration and its unique culture but also opened up entire new vistas for the field itself. The new story of the mathematician as an explorer made possible a new set of questions and legitimized new approaches that had previously been banned from the realm of proper mathematics. Specifically, the infinitesimal techniques that transformed mathematics and led directly to the calculus of Newton and Leibniz were inspired and supported by the narrative of mathematical exploration and were inseparable from it.[3]

Duel at Dawn focuses on the stories that succeeded the narratives of exploration and helped guide the development of mathematics in the eighteenth and nineteenth centuries. During the Enlightenment geometers were idealized not as enterprising voyagers but as innocent children of nature, impervious to the corrupting influence of human society. They were simple "natural" men who, unlike the rest of us, had never lost their childlike curiosity about the world and were therefore uniquely suited for the task of uncovering the hidden harmonies that govern our seemingly chaotic universe.

Not coincidentally, that was precisely the role assigned to geometers during the age of the Enlightenment: abstracting from physical reality in order to reveal its deep mathematical structure. The iconic mathematician of this age was Jean d'Alembert, a foundling raised by a humble foster family, who was nevertheless a natural aristocrat who grew up to become the toast of the Paris salons. Here was a true child of nature, whose unmediated connection to the world was manifest in his brilliant mathematical insights. At the same time d'Alembert was also the chief public apologist for mathematics, insisting that it was firmly rooted in actual relations in the physical world.

The tale of the mathematician as a natural man in tune with the peaceful harmonies of the world did not survive the upheavals of the late eighteenth and early nineteenth centuries. In the early decades of the nineteenth century the new story of the tragic young genius challenged the prevailing narrative of the Enlightenment "natural man" and over time replaced it. Galois, the misunderstood outcast who was cut down in the prime of youth, replaced the celebrated and accomplished d'Alembert as a model of the true mathematical life. At the same time a new practice emerged that radically divorced mathematics from physical reality and focused instead on the creation of an alternative mathematical universe, governed solely by rigorous mathematical reasoning.

Most often, those who were depicted as tragic heroes were also among the leading champions of the new rigorous mathematics. This is true not only of young immortals like Galois and Abel but also of Augustin-Louis Cauchy, the most influential mathematician of the first half of the nineteenth century. Although Cauchy stars as the chief villain in the legends of Galois and Abel, to himself and his admirers, he too was a martyr to truth. Hounded out of France by political rivals in the wake of the revolution of 1830, he never wavered in his religious and political stance, willingly paying a heavy personal and professional price. At the same time Cauchy was the most prominent champion of the new rigorous mathematics, persisting in his advocacy in the face of determined opposition from his colleagues at the École polytechnique and the Academy of Sciences.

The "enterprising explorer," the "child of nature," and the "tragic genius" each formed a distinct story that characterized mathematics and its practitioners at some point during the past three centuries. Each of these stories drew on cultural narratives that prevailed at the time: the heroic voyagers on the great age of exploration; the "child of nature" on the Enlightenment ideal of the uncorrupted "natural man"; and the young martyrs on the imagery of High Romanticism, which celebrated poets, artists, and musicians who shone fiercely and died young. In each case these narratives went hand in hand with a distinct mathematical practice: the paradoxical infinitesimal methods of the seventeenth century; the Enlightenment search for the mathematical harmonies of nature; and the modern rigorous mathematics, occupied with a pure mathematical universe separate from our own. In each case the stories place a distinct mathematical style and practice at the heart of a well-known cultural milieu of its time. Overall, viewing mathematics through the lens of mathematical stories reveals a secret history of the field, in which legends of hidden riches in undiscovered lands and beautiful dreamers cut down in the flower of their youth play as great a role as Euclidean geometry and the differential calculus. Through it all, mathematics, even in its highly abstract forms, becomes an inseparable part of human culture and human history.

This book's main theme covers a period of nearly two centuries in the history of mathematics, from early work on the calculus in the late seventeenth century to the establishment of the "modern" style of mathematics in the nineteenth century. Part I, "Natural Men," deals with the mathematical stories and mathematical practice of the Enlightenment, from the 1690s to around 1800. Chapter 1 follows the life and career of Jean le Rond d'Alembert, one of the leading geometers of the age, coeditor of the *Encyclopédie,* and a cultural hero of the Enlightenment. Aided by the remarkable story of his birth and upbringing and drawing on the pastoral ideals of the Enlightenment, d'Alembert fashioned himself as a "child of nature," unspoiled by the corrupting influence of society. This made him uniquely suited for what d'Alembert insisted was the primary task of the geometer: uncovering the hidden harmonies that govern our seemingly chaotic world. Although d'Alembert was the iconic mathematician of the Enlightenment, other less-known figures shared both his self-presentation and his views on the nature of mathematics: in the Enlightenment, mathematics was derived from nature, and mathematicians were "natural men."

The theory and practice of mathematics practiced by the natural men of the eighteenth century is the subject of Chapter 2. Along with Chapter 7, this chapter includes several relatively technical mathematical expositions, taken from the work of eighteenth- and nineteenth-century mathematicians. Readers who do not wish to engage with the mathematical examples can skip these segments without missing out on the narrative or the main argument. The chapter begins with a discussion of the Enlightenment theory of mathematics, especially as expressed in d'Alembert's "Preliminary Discourse" to the *Encyclopédie,* which emphasized the interconnectedness between mathematics and the physical world. Although few geometers wrote directly about the nature of the field, their actual mathematical practice indicates that they shared the fundamentals of d'Alembert's view. The chapter then moves to examples from the mathematical work of Enlightenment luminaries that highlight the evolving nature of the field, from the geometricist work of Johann Bernoulli in the 1690s to the highly abstract style favored by Lagrange in the late eighteenth century, with d'Alembert's and Euler's work in between. But even in its most abstract form practiced by Lagrange, Enlightenment mathematics never lost touch with its roots in the physical world. Severing the ties between mathematics and reality was tantamount to depriving mathematics of its subject matter and was therefore unthinkable.

Part II, "Heroes and Martyrs," takes us from the world of the great geometers of the Enlightenment to the very different mathematical universe that took shape in the early decades of the nineteenth century. Within a decade of the death of Lagrange in 1813 a new breed of mathematicians with a very different public persona, a different understanding of the field, and different mathematical practices had begun to challenge the "natural mathematics" of the Enlightenment. It was in those years that the image of the ideal mathematician was transformed from the tranquil persona of the natural man to that of the troubled mathematical martyr. The three chapters of this part chronicle the sudden emergence of this new mathematical story and its diffusion over the following century.

Évariste Galois, the most extraordinary of the mathematical martyrs, is the focus of Chapter 3, which traces the origins of his legend and its subsequent fortunes. A detailed study of his life and career shows that Galois was far from being the innocent victim of legend, and that many of young Galois' troubles were due to his own paranoid and provocative behavior. This, however, made no difference to his lasting reputation, for the historical Galois was quickly swept away by the rising tide of his legend. In this version Galois was a true innocent, victimized and persecuted by a faceless establishment that refused to grant him the recognition he deserved. Originating in his lifetime in the small circle of Galois and his friends, the legend was originally confined to the radical fringe of French society. By the late nineteenth century, however, Galois was firmly installed in the pantheon of French scientific heroes, and by the early decades of the twentieth century his fame had spread to the English-speaking world as well. The marginal young man who was left to die on a Paris street had come to represent the ideal of a true mathematician.

Niels Henrik Abel, the subject of Chapter 4, was a very different young man from his younger contemporary, Galois. Amiable and moderate where Galois was prickly and radical, young Abel was making a name for himself in the 1820s in the highest mathematical circles. Lacking a regular academic position in his Norwegian homeland, and with several siblings dependent solely on his income, Abel's life was a difficult one. But his mathematical star was on the rise, and with influential friends working on his behalf, he had a bright future ahead of him when he contracted tuberculosis and died at the age of 26. Remarkably, the real Abel, the conventional young man with career aspirations, was immediately forgotten at the moment of his death, replaced by a very different character: an innocent abroad, hopelessly unfit to care for himself in a cruel world and driven to his death by vindictive, petty men. In his death Abel became Galois' twin, a saintly mathematician persecuted by the faceless "powers that be." Although in truth neither mathematician actually fit this mold, the prevailing legend proved stronger than memory and forced their biographies into a standard romantic narrative. By the early twentieth century both young men had become ideal examples of the pure mathematical life. Augustin-Louis Cauchy, the subject of Chapter 5, was the leading mathematician of his day, a member of the Paris Academy of Sciences and holder of numerous positions in the most prestigious centers of learning in Europe. He was also cited as the chief villain in the legends of Galois and Abel and therefore seems an unlikely candidate for the role of a persecuted martyr. Nevertheless, in his own eyes and those of his friends and admirers, that is precisely what he was. As an outspoken Catholic and royalist, Cauchy was often vilified by his scientific colleagues, most of whom remained loyal to the republican ideals of the Revolution. As an uncompromising mathematical reformer and advocate of a new and rigorous approach to the field, he conducted a decade-long battle with his superiors at the École polytechnique, who insisted that he teach the curriculum, not his own singular views. As long as the conservative Bourbons were in power, they protected Cauchy from the attacks of his enemies, but when the political winds shifted, he went into exile and quickly lost most of his positions and honors. Undeterred, Cauchy remained as uncompromising as ever, standing up for political, religious, and mathematical truth and willingly paying the price for his convictions. In his own eyes and those of his admirers, Cauchy, like Galois, was a martyr to the truth.

Part III, "Romantic Mathematics," focuses on the ideal and the practice of the new mathematics that emerged in the early nineteenth century. Chapter 6 locates the new persona of the tragic mathematician within the context of the ideals of High Romanticism and the popular stories told at that time about poets, artists, and musicians. Unmistakably, the tales told of Galois and Abel are of a kind with those told of Byron, Keats, Chopin, and (later) Van Gogh. Not coincidentally, the new mathematicians and the romantic artists and poets also shared an understanding of their role in the world and the meaning of their craft. Broadly speaking, romantic artists and poets rejected the Enlightenment's rational and analytic approach to the world and sought to connect with deeper emotional levels and higher realms of the sublime. Similarly, the new mathematicians turned away from the Enlightenment focus on analyzing the natural world to create their own higher reality—a land of truth and beauty governed solely by the purest mathematical laws. In both cases the practitioner, whether poet, artist, or mathematician, could no longer be the "natural man" of the Enlightenment, who was uniquely attuned to the rhythms and harmonies of nature. He was, rather, an otherworldly creature who belonged in a better and truer world than our own. Obliged to spend his life in our deeply flawed world, the romantic hero inevitably succumbed, falling victim to the machinations of lesser men.

Thus the new mathematical persona and the new mathematical practice are inseparable from each other. A mathematics that seeks to build its own pristine universe requires an "otherworldly" mathematician who can soar into higher realms and will never belong in our mundane material world. At the same time the persistent story of the tragic young mathematical genius suggests a mathematical world so different and separate from ours as to make its denizens strangers in our own world. The story of the tragic genius and the insular practice of the discipline: each requires the other, and both are inseparable from the romantic movement that peaked in the same years.

Chapter 7 examines the technical work produced by the mathematical poets. As with Chapter 2, readers who do not wish to engage with the more technical mathematics can skip these sections without missing out on the narrative and the main argument. Cauchy broke with the Enlightenment tradition by reformulating the calculus as a self-contained rigorous system, divorced from the material world and based on purely mathematical concepts. Galois resolved one of the long-standing problems in the theory of equations through a profound investigation of the field, leading to a general solution that he himself proclaimed was completely useless. Future mathematicians, Galois argued presciently, would dispense with searching for useful solutions and focus instead on the deep architecture of mathematics itself. In due time, he predicted, mathematics would be freed of all external requirements and become its own subject matter. Overall, at the apex of the age of Romanticism mathematics broke away from the natural world and the natural sciences. In its narratives, as well as in its practice, mathematics fashioned itself as a creative art, devoted to the pursuit of sublime truth.

Part IV, "A New and Different World," consists of a single chapter and revisits the book's argument from the perspective of one of the most spectacular achievements of nineteenth-century mathematics: the invention of non-Euclidean geometry. For more than two millennia Euclidean geometry had been not only the foundation of all mathematical knowledge but also a universal model for true knowledge in all fields. Since ancient times Euclid's opus had served as an unassailable rock of certainty in the turbulent waters of contested human knowledge. Only geometry, it seemed, was free of the perpetual disputes that plagued all other intellectual pursuits; it compelled assent and provided true, unchallengeable knowledge about the world. Placed at the core of all mathematical work, it also served as an anchor for mathematics, connecting even the most abstract work to its origins in the physical world. All this, however, changed with stunning rapidity in the early decades of the nineteenth century. Previous efforts to correct a perceived flaw in Euclid's edifice were transformed into the complete abandonment of Euclidean geometry as a model of fixed, unassailable truth. In place of a single true and necessary geometry there arose a multitude of other non-Euclidean geometries, clearly at odds with our worldly experience but as logical and coherent as Euclid's own.

More than any other mathematical achievement, non-Euclidean geometry embodies the profound transformation in the character and understanding of the field that took place in the nineteenth century. Long viewed as an expression of the deep relations that prevail in our world, mathematics was unmoored from its foundations in physical reality and cast adrift in conceptual space. As long as it was logically sound and answered to the highest mathematical standards, a non-Euclidean world was as true and as real as a Euclidean one. The fact that it conflicted with our own experience of the physical world mattered not at all.

The chapter follows the careers of several innovators in the field but focuses on János Bolyai, who as a young officer in the Austrian army developed a complete non-Euclidean system of geometry. Bolyai's story, however, was a tragic one: denied the recognition he deserved, he spent his remaining years in bitter retirement, never again contributing to the field he helped transform. In his own eyes and those of his admirers in later years, Bolyai was a tragic romantic hero, devoted to the pursuit of truth and beauty but ignored and misunderstood by his fellow men. He was a man in the mold of Lord Byron or Van Gogh, but also of his mathematical contemporaries Galois, Abel, and Cauchy. For all of them, the vision of a self-contained mathematical universe freed from our material world was inseparable from a vision of the mathematician as a seeker of sublime truth, at home in the rarefied world of mathematics but unsuited for life in the crass world of men.

The Conclusion, titled "Portrait of a Mathematician," restates the book's general argument in visual terms. Portraits of the tragic mathematical heroes of the early nineteenth century show an unmistakable affinity to those of the poets and artists of the age. These romantic images depict intense young men with blazing eyes, focused not on us but on greater truths beyond the horizon. The portraits of these mathematicians are very different, however, from portraits of the great geometers of the Enlightenment, as well as from those of nineteenth-century scientists. These depict active and engaged men in the prime of life, directly addressing and engaging the viewer. In their visual representations, as well as in their legends and their mathematical practice, Galois and his contemporaries broke away from the traditions of the Enlightenment and from their close association with the natural sciences. Whereas the grands géomètres of the Enlightenment were presented as worldly, successful men, the new mathematicians were depicted as romantic strivers for sublime perfection, ill at ease in their mundane surroundings.

The Conclusion notes that like its predecessors, the modern style of mathematics will also in time give way to a new narrative and a new type of practice. One possible direction is suggested by the ongoing debate among academic mathematicians over the role of computers. Most mathematicians today oppose increased reliance on computers, but the move is supported by a dynamic group of advocates. If proof by computer does ultimately gain legitimacy in the mathematical community, one can expect significant changes in the narrative and imagery associated with the field as well. It is possible, for example, that the stereotype of the power-hungry and resentful computer "wiz" may encroach on the saintly image of the mathematical martyr.

In the end, however, there is no telling which way the field will turn. Just as the geometers of the Enlightenment could not imagine the mathematical practice and the surprising ideal of the mathematical practitioner that would succeed them, so we too are in the dark about what lies ahead. Mathematics may turn toward the power of computers, or it may develop in a wholly different direction that is quite unthinkable to us now. The only thing that is certain is that mathematics will be transformed once again, as it has been repeatedly in the past. And as mathematics changes, its stories, legends, and images will change along with it.

I

NATURAL MEN

1

THE ETERNAL CHILD

The Foundling

On the night of November 16, 1717, a policeman on his rounds came across a small wooden box on the steps of the church of St. Jean le Rond in Paris. Upon closer inspection he found in it a newborn baby boy, exposed to the elements and the mercy of strangers. Foundlings were not rare in eighteenth-century Paris, and most ended up in notorious publicly run orphanages from which only the hardiest ever emerged. But the baby found that night was more fortunate: he was treated with care and kindness and grew up to become one of the shining stars of the Enlightenment. He was Jean le Rond d'Alembert, a man who spent his days in the sparkling salons of Paris, but whose middle name forever harkened back to the infant who was abandoned on the steps of a church on a cold November night.

The mystery surrounding d'Alembert's parentage did not last long. Within a few days of his discovery it became known that the foundling was the illegitimate son of the scandalous Mme. de Tencin and one of her lovers, the Chevalier Louis-Camus Destouches (1668–1726), officer of the artillery. The daughter of the president of the parlement of Grenoble, Claudine Alexandrine Guérin de Tencin (1682–1749) was brought up to be a nun, but she obtained a release from her vows from the pope. She moved to Paris, where she established one of the first fashionable salons, a model for the prestigious social institutions in which her son sparkled a generation later. She played hostess to philosophers and writers and eventually published several novels under the names of her nephews. She was well connected at the courts of Louis XIV and Louis XV and is widely credited with having her brother appointed a cardinal. But more than anything Mme. de Tencin was famous for her many love affairs, which in one instance earned her a jail cell in the Châtelet, following the suicide of one of her lovers. When her liaison with the Chevalier Destouches resulted in a pregnancy, it was for her merely an inconvenient distraction. She abandoned the baby to the mercies of strangers and continued her active social and political life as before. Although she lived till d'Alembert was well into his thirties and a rising star in Parisian society, Mme. de Tencin never acknowledged the son she had left on the church steps.

Not so d'Alembert's father, the chevalier. Although he was away from Paris at the time of d'Alembert's birth, Destouches sought him out as soon as he learned of his existence and took charge of his upbringing. He placed d'Alembert in the care of a glazier's wife, Mme. Rousseau, who raised him with the loving care of a mother and provided him with a family refuge in later years. Indeed, so close did d'Alembert and his foster mother become that he remained in her humble home for nearly half a century before moving out in 1765. The Destouches family meanwhile continued to provide for d'Alembert, and when the chevalier died in 1726, he left his son a comfortable pension to live on. He also made sure that his family would continue to care for his illegitimate son and provide him with a proper gentlemanly education. At age 12, thanks to the patronage of the Destouches family, d'Alembert entered the prestigious Collège des quatre nations (also known as the Collège Mazarin) as a "gentilhomme" and spent his next six years there.[1]

D'Alembert's early years, then, were comfortable, supported by the warmth of a loving foster family and the resources of an aristocratic clan. Nevertheless, both he and his admirers, as well as many later biographical notices, made a point of emphasizing that this great man of letters and science, a giant to his contemporaries, began life in the humblest of circumstances, as a foundling abandoned at the door of a church.[2] Even though his parentage was well known to anyone who sought to learn it, he yet viewed and presented himself as an abandoned orphan, bereft of family and connections, making his way in the world on the strength of his merit alone.

The events surrounding d'Alembert's birth were recounted many years later, after d'Alembert's death in 1783, by his friend and protégé Marie-Jean de Caritat, Marquis de Condorcet (1743–1794). Thanks in part to d'Alembert's patronage, Condorcet was at that time the perpetual secretary of the Paris Academy of Sciences, where one of his chief duties was to author official eulogies for members of the academy upon their passing.[3] Condorcet was devoted to the memory of his mentor and sought to preserve his legacy and a touch of his personality for future generations. As a result, his "Éloge de M. d'Alembert" was one of the longest and most personal eulogies he ever wrote.[4]

Although Condorcet undoubtedly knew all there was to know about d'Alembert's family and connections, he nevertheless made much of his supposedly obscure parentage. "We will not seek to lift the veil which concealed the names of his parents during his lifetime," Condorcet wrote at the beginning of the "Éloge." "What importance could their identity have? The true ancestors of a man of genius are the masters who had preceded him in his vocation; and the true descendants are the students worthy of him."[5] In presenting the great mathematician as belonging to the world as a whole rather than to a particular clan, Condorcet was making an ideological point: the worth of a true philosophe such as d'Alembert arose not from family connections or aristocratic birth, but from personal genius and the unfettered pursuit of truth.[6]

Condorcet then recounted the basic facts of d'Alembert's early days:

> Exposed near the church of St. Jean le Rond, M. d'Alembert was brought to a police commissioner, who fortunately was not hardened by the habit of the sorrowful duties of that place; fearing that this weak and nearly dead infant would not find the care and attentions necessary for his survival in a public hospice, he placed him in the charge of a working woman whom he knew to be moral and humane; and it was upon this happy coincidence that depended the survival of a man who became the honor of his country and his age, and whom nature had destined to enrich with so many new truths the system of human knowledge.[7]

To Condorcet, it seems, d'Alembert's life followed a familiar trajectory of legend. He was the child of obscure parentage, alone in the world, who was adopted and raised by kindly strangers. Years later the child proved to be no ordinary man but a chosen one, destined to forge new paths never before trodden. The precise parentage of d'Alembert did not matter to Condorcet, for he transcended it. Like other such foundlings in history and myth, d'Alembert belonged to the world and was destined to bestow on his followers wondrous treasures of true knowledge.

If the outlines of the story sound familiar, it is, of course, because they recur time and again in the Western tradition. Moses of the Old Testament was such a foundling, as were Remus and Romulus in the Roman tradition. But the true child of mystery in the Western tradition, whose parentage was obscure but who was destined to change the course of human destiny was, of course, Christ. Young Jean le Rond, like the infant Christ, was a child whose apparent family was not his true one. Both belonged to humanity as a whole and were destined to lead men forward to salvation (Christ) or to new realms of knowledge (d'Alembert). It is the biblical story of the Nativity, and it is full of hope and promise. The infant abandoned before the church of St. Jean le Rond, Condorcet wrote, was to become "the honor of his country and his age . . . destined to enrich with so many new truths the system of human knowledge."[8]

The seeds of d'Alembert's meteoric rise were sown when he first encountered advanced mathematics as a student at the Collège des quatre nations. Mathematics was not a main topic of instruction at the college, which was a Jansenist religious school, dedicated to preparing its students for the clergy. Fortunately for d'Alembert, however, the most prominent professor in the college's history had been Pierre Varignon (1654–1722), the leading French mathematician of the previous generation. Thanks to Varignon's influence, the college offered a full year of instruction in mathematics and, even more important, 2,500 volumes devoted to the field in its library. Young d'Alembert took to mathematics with a passion, and by the time he received his *baccalauréat* in 1735, he had become aware of his uncommon talent for the field.[9] At the same time, much to the disappointment of his teachers, he also developed a strong distaste for theology and refused to consider an ecclesiastical career.

Mathematics may have been d'Alembert's true love, but then, as now, it did not promise a secure future for a young man of limited means. Feeling compelled to pursue a more practical occupation, he put aside his own intellectual interests and for two years turned to the study of law. D'Alembert earned his advocate's license in 1738 but, unlike his fellow students, declined to enter on a legal career. The pattern that had been set in his study of theology repeated itself: by the time he had completed his studies, he had become completely disenchanted with the field and could not conceive of spending his life as a lawyer.

Seeking an alternative, d'Alembert began studying medicine in 1739, but again things did not go well: "The passion for geometry caused him once more to neglect his new studies," Condorcet wrote. Seeking to concentrate on his medical studies, he decided to part with the "objects of his passion": "His mathematical books were carried off to one of his friends, from where he was not to retrieve them until he had become a

Doctor of Medicine, when they would be no more than a diversion to him, rather than a distraction."[10] Things did not go as planned, however: time after time he asked his friend to return just one book to him, to help remind him of a particular method or proof, until soon all the books were back in d'Alembert's possession. "Then, well convinced of the hopelessness of his efforts to combat his penchant for mathematics, he conceded" and abandoned his medical studies. Instead, "he devoted himself forever to mathematics, and to poverty."[11]

As it happened, d'Alembert did not have to suffer excessive hardships in his pursuit of a career in mathematics. He moved back in with his foster family, where Mme. Rousseau, "who loved him like a son," took care of all his needs.[12] There he stayed for decades, living simply in the company of working folk even as his reputation as a mathematician and man of letters soared in the outside world. Blissfully ignorant of the fame of her adopted son, Mme. Rousseau took to teasing d'Alembert for his constant preoccupation with writing pointless treatises, telling him that he was "nothing but a philosopher." When d'Alembert in return asked her what a philosopher was, she answered that "he is a fool who torments himself during his life about what will be said of him when he is no more."[13]

The self-deprecating humor evident in this story was, according to Condorcet, typical of his mentor. Indeed, in his "Éloge" Condorcet made much of the humble surroundings in which d'Alembert worked, and of the modesty and unpretentiousness with which he pursued his studies. Sometimes during these early years, Condorcet recounted, d'Alembert would pursue an idea on his own and "tasted the pleasure of making a discovery." Alas, upon further reading d'Alembert would soon discover that what he had thought was a discovery was, in fact, already known. "He was therefore persuaded that nature had refused him genius, and that he must be limited to knowing that which others had discovered." Nevertheless, Condorcet continued, d'Alembert in his modesty showed no resentment at this assessment of his accomplishments. "He felt that the pleasure of studying, even without glory, was sufficient for his happiness."[14]

The picture of d'Alembert that emerges from Condorcet's account of his early years is appealing. Here was a child left alone in the world, devoid of family or connections, living a life of simplicity itself, untouched by ambition, greed, or vanity. Rather than pursue a respectable career in the professions, the young d'Alembert chose to follow his heart and was willing to sacrifice both social status and financial rewards in this pursuit. His study of mathematics was not even based on ambition or dreams of immortal fame, because, according to Condorcet, d'Alembert did not initially consider himself a particularly gifted mathematician. Rather, his tireless work in the field was motivated solely by his true love of mathematics. Unlike Mme. Rousseau's self-involved philosopher, d'Alembert was unconcerned with his reputation, either current or posthumous. He lived entirely in the present, following his heart's desires without regard to material goods, fame, or fortune. Such purity, Condorcet observed, was the sign of a man untouched and uncorrupted by the evils of society. "It is rare," he wrote, "to be able to observe the human heart so close to its natural purity, and before amour-propre corrupts it."[15]

For Condorcet, then, d'Alembert was a pure "natural man," untouched and uncorrupted by the artificial and superficial mores of high society. His obscure parentage dissociated him from any particular social lineage, class, or background. His contentment in the humble company of his foster family exemplified his modesty and sincere disregard for wealth, social station, and earthly glory, as did his ardent pursuit of a vocation that promised very little in the way of material rewards. D'Alembert was a child of nature, and his heart retained the purity and simplicity of a child, which were lost to the rest of us.

Undoubtedly this idealized vision reflected Condorcet's conception of an ideal philosophe and man of science as much as his memory of the living d'Alembert. In truth, d'Alembert's parentage was well known to his contemporaries, and his high birth, far from being an obstacle to his pursuit of mathematics, likely played a role in opening doors that would have been closed to a true commoner. When d'Alembert abandoned promising careers in law and medicine, he was, as Condorcet emphasizes, following his heart, but he could afford to do so because his father had left him a comfortable pension to live on. This is a far cry from the "poverty" to which he allegedly resigned himself in his pursuit of his love of mathematics. Condorcet may have presented him as a "child of nature" devoid of family, but here again we see that it was, in fact, d'Alembert's well-placed family that helped him pursue a life in mathematics. Finally, Condorcet's insistence that d'Alembert cared little for earthly glory seems at the very least exaggerated if we consider the determination and skill with which he pursued an appointment to the Paris Academy of Sciences. In the years after his graduation from the Collège des quatre nations d'Alembert presented several papers to the academy on a range of mathematical topics and submitted his candidacy for membership three times in quick succession. On his third try, in 1741, he was finally elected a member of the academy.

Jean-Jacques and Jean le Rond

All this is not to deny that Condorcet, who knew d'Alembert intimately for many years, did his best in the "Éloge" to communicate to posterity his friend's personal character and charm. Undoubtedly he did, and one cannot read Condorcet's account of d'Alembert without feeling a true human presence behind the elegant prose. Nevertheless, Condorcet's "Éloge" is significant not so much for giving a fair-minded assessment of d'Alembert's character and work, but rather for presenting an idealized vision of what a mathematician should be. D'Alembert, for Condorcet, embodied the perfect mathematician, and for him, such a person was first and foremost a "natural man."

The notion that sometime in the distant past human beings had existed in a "state of nature" was a widespread hypothetical ploy among philosophers of the seventeenth and eighteenth centuries. Thomas Hobbes (1588–1679), John Locke (1632–1704), and Jean-Jacques Rousseau (1712–1778) all wrote treatises discussing this hypothetical state and describing men's transition from the state of nature to civilization.[16] The English philosophers differed radically from each other in their account of life in the natural state, Hobbes famously describing it as "nasty, brutish, and short" and Locke suggesting that it was quite rational and relatively peaceful. Both, however, viewed the emergence of civilization as a necessary and, on the whole, benevolent transition.

Not so Rousseau, who argued that men in the state of nature were naturally good and that life in civilized society undermined men's goodness and ultimately rendered them unhappy. This transformation, according to Rousseau, occurred when natural *amour de soi,* or self-love, becomes the civilized *amour-propre.* The former is a benevolent sentiment, focused on the self alone and the fulfillment of one's needs; the latter is a needy and demanding sentiment, based on comparing oneself with others. Civilized men, according to Rousseau, are guided by amour-propre and become envious, petty, and vindictive. "This is how the gentle and affectionate passions are born of self-love [*amour de soi*] and how the hateful and irascible passions are born of *amour-propre,*" Rousseau wrote. "Thus what makes man essentially good is to have few needs and compare himself little to others; what makes him essentially wicked is to have many needs and to depend very much on opinion."[17]

In his pedagogical treatise *Emile* Rousseau insisted on the paramount importance of independence from society's mores and superficial values. Emile, Rousseau's paragon of natural virtue, is brought up in isolation in his early years to prevent his tender soul from being corrupted by amour-propre. This, however, does not mean that Rousseau was advocating a withdrawal from society or a life of contemplation in the manner of medieval monastic orders. Far from it: the ultimate goal of Emile's education is to make him an active member of society and a joy to his companions, both male and female. Because of his unique upbringing and his resistance to social evils, Emile will ultimately become an outstanding member of society, far superior to those who grew up under its sway.[18]

Rousseau was not content to have his views debated among professional philosophers, or even among members of a narrow elite. Through his romantic novels and a lively personality cult of Jean-Jacques, Rousseau's ideas were disseminated to an ever-growing segment of French society. In 1783, when Condorcet wrote his "Éloge" of d'Alembert, Rousseau worship was at its height in fashionable Parisian circles, and it is in this context that we should read Condorcet's references to d'Alembert as a "natural man." For the "Éloge" is, in fact, imbued with Rousseauian sensibilities.

When he wrote of d'Alembert that "it is rare to be able to observe the human heart so close to its natural purity, and before amour-propre corrupts it," Condorcet was using explicitly Rousseauian terminology.[19] As his contemporary readers would have been well aware, Condorcet was suggesting that d'Alembert was exceptional among men for having preserved his pure and natural outlook on the world in the face of the corrupting influence of society. Untouched by amour-propre, d'Alembert was free of envy, pettiness, and ambition and could devote himself wholeheartedly to his true passion, the study of mathematics. It was this natural innocence, according to Condorcet, that allowed d'Alembert to live happily for decades in the company of working folk, "preserving always the same simplicity."[20] It was the same absence of amour-propre vanity, according to Condorcet, that sustained him when he believed himself but a mediocre and unoriginal mathematician, and enabled him to persist in his studies of his chosen field. And it was his lack of socially induced ambition that permitted him to abandon promising careers in law and medicine in order to pursue "mathematics and poverty."[21] It is hardly surprising, then, that in Condorcet's account the early years that followed his resolution to pursue mathematics were the happiest of d'Alembert's life.[22] For it was in those years that he was a true Rousseauian hero, insulated from the demands and expectations of society and free to follow the pure inclinations of his heart.

The Grand Geometer

The early years of poverty and isolation may have appeared to the aging d'Alembert to have been the happiest of his life, but it is worth noting that at the time he made no effort to prolong them. They came to an abrupt end when, after an intense campaign, d'Alembert was elected a member of the Academy of Sciences in 1741. It was hardly a year later that d'Alembert became embroiled in a fierce priority dispute with fellow academician Alexis Clairaut (1713–1765)—precisely the kind of quarrel fueled by pettiness and envy that Rousseau had warned against. The dispute arose when in late 1742 d'Alembert began reading to the academy chapters of his *Traité de dynamique,* a work that was to become one of the most important in his career. Within a month, while d'Alembert was still in the midst of his recitations, Clairaut began reading his own paper on the principles of dynamics. Alarmed that he might be upstaged and robbed of credit, d'Alembert responded in a manner altogether unbecoming of one supposedly lacking in amour-propre and ambition. In order to avoid the long wait for the academy's official publications, he rushed the treatise to print that very year, before it was fully ready. The result is a work that has been much praised for its insights, but also much criticized for being obscure and unreadable. The episode also earned d'Alembert the long-standing enmity of Clairaut, who had previously been a strong supporter of his younger colleague.

The rivalry between d'Alembert and Clairaut continued to flare up repeatedly until the latter's death in 1765. Time and time again the two wound up working on the same or closely related topics, competing with each other for credit and acclaim. Clairaut became so frustrated with his repeated clashes with d'Alembert that a year before his death he complained to his Jesuit colleague Ruggero Boscovich (1711–1787) of d'Alembert's "fanatical eagerness" to take on the same subjects as himself.[23]

But Clairaut was not the only mathematician with whom d'Alembert carried on a lifelong rivalry. Like the other *grands géomètres* (to use d'Alembert's term) of the age, d'Alembert lived in constant anxiety that his accomplishments would be overshadowed by the great Swiss mathematician Leonhard Euler, whose superior analytical abilities were matched by the sheer volume of mathematical works streaming steadily from his pen.[24] The relationship began promisingly enough when in 1746 d'Alembert won an essay competition on the theory of the winds, sponsored by the Royal Academy of Sciences and Belle-Lettres in Berlin where Euler was the resident mathematician.[25] By the late 1740s, however, the two became the chief antagonists in a controversy over the mathematical theory of vibrating strings, which drew in all the leading geometers in Europe. The debate soon evolved into a general consideration of the nature of functions, with d'Alembert promoting a more restrictive interpretation of what constitutes a function than Euler. Although the issue was never actually settled, d'Alembert eventually became a lone holdout for his interpretation, while the vast majority of his colleagues sided with Euler.[26]

Tensions increased in the early 1750s, after d'Alembert published treatises on the precession of the equinoxes (the wobble of the earth's axis) and the theory of the resistance of fluids.[27] As was often the case with d'Alembert's works, the treatises contained important and original contributions presented in a haphazard and barely comprehensible manner. In both cases Euler adopted d'Alembert's general approach but gave the topics a far clearer and more thorough treatment than was available in the original. Startled, d'Alembert wrote a number of sharply worded letters to the Berlin Academy demanding that Euler clearly and publicly acknowledge d'Alembert's priority. Euler did so, publishing a note in the Berlin Academy's journal stating that he "did not make the slightest pretension to the glory that was due" to d'Alembert. But the seeds of discord were sown, and suspicion between the two men lasted for years.

Things got worse before they got better. Although Euler was the leading savant of the Berlin Academy, he was far from satisfied with his position there. Unlike the Paris Academy of Sciences and the Royal Society of London, which enjoyed a great deal of autonomy, the Berlin Academy was very much the creature of its royal patron, King Frederick II of Prussia, who took an active interest in its daily affairs. Success in the Berlin Academy depended largely on finding favor with Frederick, and unfortunately for Euler, the monarch took no liking to the Swiss mathematician.

Frederick was a patron and even a friend to some of the leading geometers of his era, so it is surprising to note that he was, in fact, completely indifferent to mathematics. The study of mathematics "dries up the mind," he once complained in a letter to his friend Voltaire.[28] "An Algebraist, who lives locked up in his cabinet," Frederick wrote to d'Alembert, "sees nothing but numbers and propositions, which produce no effect on the moral world. The progress of manners is of more worth to society than all the calculations of Newton."[29] The king was, however, deeply enamored of French culture, wit, and belles lettres and was always intent on increasing French membership in his academy. In 1746, after years of courtship, he managed to lure the Parisian luminary Pierre-Louis Moreau de Maupertuis (1698–1759) to Berlin to become president of the academy by asking him to show "to a King how sweet it is to possess such a man as you."[30]

But Frederick had little patience for the humorless German-speaking Euler, who was all business and had no talent for courtly witticisms. In his letters the king called the half-blind Euler "a huge cyclops mathematician" and mocked his clumsy efforts to play the elegant courtier.[31] Because Euler was unquestionably the shining scientific star of the academy, Frederick was not inclined to let him go. Instead, he loaded him with administrative duties and never showed him the respect due to his mathematical reputation. When Euler finally left Berlin in 1766 to return to his former position in the Imperial Academy of Sciences of St. Petersburg, Russia, Frederick wrote scathingly to d'Alembert:

> Mr. Euler, who loved the big and the little bears unto madness, has gone to the north to observe them more at his ease. A ship which carried his *xz*'s and *kk*'s has been wrecked; all has been lost, and it is too bad, because he would have had enough to fill six volumes in folio of memoirs full of ciphers from one end to the other and Europe will be truly deprived of the agreeable amusement that this reading would have provided.[32]

The fact that Frederick chose d'Alembert as the recipient of these unkind witticisms was no coincidence, for while Frederick considered Euler an insufferable bore, he showed nothing but admiration to his French rival. D'Alembert, in Frederick's eyes, was everything that Euler was not: sociable, witty, and a man of letters, whose scientific genius was fully matched by his reputation as a writer and philosophe. He was also, of course, a Frenchman and consummate Parisian, both of which represented to Frederick the glamour of high society and the epitome of culture and enlightenment. When Maupertuis left Berlin in 1756, Frederick did not wait long before inviting d'Alembert to take his place as president of the academy.

As one would expect, Frederick's courtship of d'Alembert irritated Euler no end. He was still the most prominent member of the Berlin Academy, and following Maupertuis' departure he naturally became the institution's acting president. Frederick, however, never intended to confirm him in this position, and although Euler was responsible for fulfilling all of Maupertuis' former duties, he received few of the benefits and honors that had accrued to his predecessor. The king, meanwhile, was showering d'Alembert with gifts and even provided him with a pension, all with the aim of luring him to Berlin.

D'Alembert, however, would not come. Paris was his beloved home, and he had no desire to forsake it for what he considered an uncouth city in the barbaric east. He was, furthermore, concerned about the effect that overdependence on a great prince could have on a man of letters such as himself. In his *Essai sur la société des gens des lettres et des grands* (1753) he warned against the dangers of making the *grands* into arbiters of arts and letters and recommended that men of letters live a life of "liberty, truth, and poverty," far away from the centers of power.[33] As we have seen, that was precisely the kind of life that Condorcet attributed to d'Alembert in his "Éloge" decades later.

D'Alembert's reasons for declining Frederick's repeated invitations seem to have been both genuine and principled, but Euler did not see them as such. Convinced that d'Alembert was intent on undermining his position, he viewed the Frenchman's reluctance to come to Berlin as a ploy to obtain a larger stipend from Frederick. From the mid-1750s Euler became openly hostile to d'Alembert, refusing to consider his entry for a Berlin Academy prize competition and blocking his access to the academy's journal, the *Histoire de l'Académie Royale des Sciences et des Belles-Lettres.* This last was a particularly severe blow because, thanks in large part to Euler, the Berlin *Histoire* had become the leading scientific journal in Europe, and d'Alembert had long preferred it to the Paris Academy's *Mémoires de l'Académie Royale des Sciences.* With no official outlet for his work, d'Alembert resorted to publishing his mathematical contributions on his own in a series of *Opuscules mathématiques.*

Matters came to a head in 1764 when d'Alembert, at Frederick's invitation, arrived for a visit in Berlin. For Frederick it was but the latest round of his decade-long effort to persuade the Frenchman to become president of his academy, and Euler was convinced that d'Alembert would finally take up the offer.[34] Refusing to stay and serve under d'Alembert's leadership, Euler began making intense preparations to return to his former position in St. Petersburg. But when d'Alembert arrived, everything changed, seemingly overnight. Far from accepting Frederick's invitation, d'Alembert made it clear right away that he had no intention of moving to Berlin. Instead, he courted Euler, lavishing admiration and praise on him before Frederick and even pressing the king to confirm his appointment as president of the academy.

The rift between Frederick and Euler was by this time too deep and fraught with bitterness to be healed by d'Alembert's intervention. Euler left for Russia within two years and was replaced in Berlin by Joseph-Louis Lagrange, a Frenchman, to be sure, but one who was hardly the sparkling society wit the king had hoped for. Furthermore, as Euler's true mathematical heir, Lagrange brought with him his own heavy load of "*xz*'s and *kk*'s" and set about increasing it year by year. "M. de la Grange calculates, calculates and calculates, curves as many as you please," the exasperated king grumbled in a 1782 letter to d'Alembert.[35] But the long-standing quarrel between d'Alembert and Euler had indeed come to an end thanks to d'Alembert's generosity during his visit to Berlin. Although they never met again, the two remained on good terms and made a point of expressing their mutual admiration until their deaths in the same year, 1783.

This rapprochement was undoubtedly aided by the fact that d'Alembert in his later years was spending less and less time on his beloved mathematics and devoting most of his energies to his work as secretary of the Académie francaise and his public role as a leading philosophe. Euler meanwhile continued publishing new mathematical work well into his seventies, even after he had become completely blind, making him easily the most prolific mathematician of the century. With their lives and careers diverging, it is likely that d'Alembert felt that he need no longer compete with his great rival on purely mathematical grounds. He could now afford to give Euler full credit as the greatest of the grands géomètres of the age.

D'Alembert had indeed acted generously toward Euler when his own mathematical career was waning, a fact that gives credence to Condorcet's view of him as innately kind and magnanimous. Nevertheless, there is no denying that during his mathematical prime d'Alembert was known as quarrelsome and even petty. In addition to his decades-long battle with Clairaut and his 15-year rift with Euler, d'Alembert was also embroiled in a long dispute with Daniel Bernoulli (1700–1782). It began with a disagreement over their respective theories of fluid dynamics and continued when d'Alembert won the Berlin Academy's essay competition on the theory of winds in 1746 over Bernoulli's entry. The two later took opposing positions in the controversy over the theory of vibrating strings, and things were not improved when the Swiss Bernoulli became the confidante of his countryman Euler during the latter's rift with d'Alembert. But whereas Euler ultimately patched up his differences with d'Alembert, this was not the case for Bernoulli, who remained on chilly terms with the Frenchman for the rest of his life.

Overall, then, at the height of his mathematical career d'Alembert had prolonged and bitter disputes with Clairaut in Paris, Euler in Berlin, and Daniel Bernoulli in Basel. Because these three were, along with d'Alembert himself, the leading mathematicians of his generation, it is fair to say that he managed to alienate all of his most prominent colleagues. As is often the case with scientific rivalries, the issues at stake were priority and credit, public honors, and prestige in the scientific community. These are hardly the types of issues we would expect to occupy a man who lived in a state of "natural purity" before amour-propre corrupted him. They are, in fact, precisely the kinds of concerns that Rousseau saw as arising from the corrupting influence of society. The pursuit of honor, credit, and prestige—the source of much of d'Alembert's testiness toward his colleagues—was, according to Rousseau, a clear manifestation of the insidious effects of amour-propre.

D'Alembert, it should be noted, was no worse than his colleagues, for the great mathematicians of the Enlightenment were a quarrelsome group, quick to take offense and forever looking with a jaded eye at their closest colleagues. But unlike other members of this exclusive club, d'Alembert was known in the Republic of Letters as much more than a brilliant geometer. He was also, with Denis Diderot, editor of the *Encyclopédie,* a leading philosophe and man of letters, a member of the Académie francaise (not to be confused with the Academy of Sciences), and from 1772 its permanent secretary. Significantly, d'Alembert's reputation in this broader social and intellectual world was very different from the way he was known in the narrow world of Enlightenment high mathematics. In place of the touchy mathematician, forever looking to defend his claims against the encroachment of rivals, we find a cultured and cultivated man of letters, the most sought-after guest in Paris, beloved and admired by his peers.

A Man of the World

D'Alembert first made his mark on Parisian intellectual life in the 1740s, shortly after his election to the Academy of Sciences, when he was invited to attend the salon of Madame Du Deffand. Taking her cue from Madame de Tencin, d'Alembert's estranged mother, the Marquise Du Deffand had established a sparkling social salon in her apartment at the convent of St. Joseph. Several times a week she hosted members of the Parisian aristocratic, political, and literary elite, engaging them in conversation on the current topics of the day. As a young mathematician only recently admitted to the academy, d'Alembert would seem to be an odd choice to be invited into such an exclusive club. Not only was he quite unknown, but his professional field, mathematics, was hardly a popular one among Mme. Du Deffand's aristocratic guests, many of whom likely shared King Frederick's view that "it dries up the mind."[36]

We do not know who introduced d'Alembert to the salon, but it is quite reasonable to suppose that it was Maupertuis, his patron at the academy.[37] Years before d'Alembert, Maupertuis was already in the privileged position of being both a highly respected mathematician and a prominent presence on the Paris social scene. He was therefore well placed to bring his junior colleague into salon circles. D'Alembert may have had other entries as well into Mme. Du Deffand's circle. As the unacknowledged son of the scandalous Mme. de Tencin, he was undoubtedly an object of curiosity in the salons of Paris. His high birth, as well as his gentlemanly upbringing by the Destouches family, furthermore, gave him aristocratic credentials that counted for much in Mme. Du Deffand's circle. But regardless of the means of his introduction, young d'Alembert became an instant social success in his own right.

Within a few years of becoming a regular guest at Mme. Du Deffand's salon, d'Alembert was invited to join the rival establishment run by Marie-Thérèse Rodet Geoffrin. By reputation the tone of conversation at Mme. Geoffrin's salon was serious, and the guests were drawn from literary and philosophical circles, whereas Mme. Du Deffand's salon was a meeting place for Paris's sparkling aristocratic set. D'Alembert moved with ease between the two establishments, soon becoming the salons' most sought-after guest and an intimate friend of both hostesses.

Unlike d'Alembert's mathematical career, which can be traced in detail through his numerous publications, his active social life is far more difficult to reconstruct. No written record can ever recover the charm, wit, bearing, and humor that made the difference between brilliant success and dismal failure in the Parisian social scene. All we can do is rely on the testimony of those who met d'Alembert, observed him in the salons, and were impressed by his personal charisma. According to Friedrich Melchior von Grimm (1723–1807), whose correspondence chronicles the social life of Paris in intimate detail, d'Alembert brought to his conversation "an almost inexhaustible source of ideas, anecdotes and curious recollections . . . He speaks very well, tells stories with much precision, and makes the humor bubble forth with a grace and nimbleness that is uniquely his."[38]

Even those who were not personally sympathetic to d'Alembert could not help but be impressed by his personal charm and ease in society. The Swedish astronomer Johann Lexell was an acquaintance of Euler and shared with him a distaste for French courtly wit. Nevertheless, having met d'Alembert in 1781, Lexell could not but admit to a reluctant admiration of d'Alembert's social charm. "He recounts his stories very well and fills them with instructive and amusing anecdotes. However, he never affects this wit, that is, he does not work at it with anxiety, but he really has it."[39] Evidently, at the age of 64 d'Alembert was as entertaining, humorous, and fascinating as he had been in his younger days in the social circles where he felt most at home.

Condorcet, not surprisingly, attributed his mentor's remarkable social success to his naturalness and innocence. D'Alembert, he wrote, "preserved his natural gaiety in all its youthful innocence [*naïveté*]."[40] What made d'Alembert so engaging in the salons, according to Condorcet, was the authentic simplicity of a natural man, untouched by the corrupting influence of amour-propre. Like Rousseau's Emile, he cared nothing for the adulation of society and as a result was rewarded with even greater admiration. D'Alembert, according to Condorcet, was self-sufficient, was content in his lot, and had retained a natural innocence and curiosity about him that the cultivated set found irresistible. The Duchesse de Chaulnes, who knew d'Alembert at the height of his social success, shared Condorcet's view on the source of his charm: d'Alembert, she said, was "only a child who lived in eternal infancy."[41]

D'Alembert's meteoric rise in Parisian social circles was no small matter in the cultural and political world of the ancien régime. In the absence of an institutional public sphere, many of the discussions and decisions that we would consider a matter of public policy took place in private, in personal interactions between prominent people who knew each other well. The great salons in which members of the political and cultural elites mingled and socialized were centers of power where much of the work of governance was done.

Being women, the hostesses of these gatherings could hold no official position, but their influence was nonetheless great. By merely inviting a young aristocrat or aspiring philosophe to their gatherings, they could raise his stock enormously, just as they could shatter one's reputation simply by excluding him from their circle. Through their power as *salonnières* and their personal acquaintance with the leading men of the realm, they could make or break careers and help determine the direction of public policy. As a very prominent member of the salon milieu and the personal confidante of its leading ladies, d'Alembert was doing far more than enjoying himself in the company of friends. He was, in fact, becoming a prominent public figure, with power to help shape French cultural and political life.

D'Alembert became a leading presence in the French cultural wars almost by accident when in 1746 he was hired, along with his friend Diderot, to oversee the translation from English of Ephraim Chambers's *Cyclopaedia* of 1728. The project was modest in its scope, and the general editor of the French edition clearly intended to complete the job in a year or two. But with Diderot and d'Alembert at the helm, the work took on a life of its own. The English original was soon forgotten, and the simple work of translation turned into the *Encyclopédie*—one of the largest publishing enterprises in history and the most lasting emblem of the cultural movement that became known as the Enlightenment. All the leading lights of French scientific and literary life contributed essays to the project, including Voltaire, Rousseau, Montesquieu, Anne-Robert-Jacques Turgot, and Étienne Bonnet de Condillac (1715–1780).

D'Alembert's most lasting imprint on the *Encyclopédie* venture was not found in the 1,309 entries he contributed but in his introduction to the first volume, which appeared in 1751. The "Preliminary Discourse to the *Encyclopédie*" was a manifesto for the French philosophes, who sought to establish all human knowledge on a rational basis.[42] It became an instant classic, making d'Alembert a public figure in a way his mathematical articles never could. It led directly to his election to the Académie française, whose "40 immortals" were charged with preserving and enhancing French language and culture. When in 1772 he became the French Academy's permanent secretary, he made full use of his power to engineer the elections of his friends and fellow philosophes to the institution and did his best to shape the direction of the Paris Academy of Sciences as well.[43]

D'Alembert's rising prominence, however, eventually put an end to his partnership with Diderot in the *Encyclopédie* project. Unlike his colleague, Diderot was never considered a respectable denizen of polite society, was not welcome in the leading salons, and was never elected to either of the two Parisian academies. He moved in more radical circles, socialized in bohemian cafés rather than salons, and never shied away from offending men in high places. Over time it also became clear that the philosophical differences separating the two editors of the *Encyclopédie* were considerable. D'Alembert, despite lavish tributes to English empiricists such as Bacon and Locke in the "Preliminary Discourse," was a rationalist who believed that physical phenomena were the expressions of underlying mathematical relations. Diderot, although initially sympathetic to this position, came increasingly to identify with the views of Paul-Henri Thiry, Baron d'Holbach (1723–1789), his friend and the leading materialist in France. Accordingly, Diderot came to believe that all matter was imbued with a vital force all its own, which could never be subjected to elegant mathematical laws or even be described by them. His *La rêve de d'Alembert* (D'Alembert's Dream), written in 1769, laid out his materialist views while presenting d'Alembert as a good-natured bumbler, disconnected from reality. Not coincidentally, it was not published until long after both protagonists were dead.[44]

Despite all this, the two editors of the *Encyclopédie* worked closely together for over a decade. When in 1749 Diderot was thrown into Vincennes prison following the publication of his *Lettre sur les aveugles* (Letter on the Blind), d'Alembert stood by him and helped secure his release. When in 1752 the enemies of the philosophes, now known as the *devots,* forced the suspension of the *Encyclopédie*'s license, the two editors worked together until the license was restored. But six years later d'Alembert had had enough. The attacks on the project by the devots were increasing, and he himself had become a favorite target. His philosophical differences with Diderot had frayed their friendship, and he had come to believe that the *Encyclopédie* was no longer the best venue for advancing the philosophes' agenda.

The project had become, in his view, too much of a flash point within the cultural and political establishment and too vulnerable to the attacks of its enemies. It was preferable, he told Voltaire, to increase gradually the membership and influence of their allies within the academies until these traditional centers of French cultural life would be in philosophe hands. Thus, after 12 years that saw both triumph and disaster, d'Alembert ended his involvement with the project that had made him a public figure and a leader of the cause of the Enlightenment. On January 1758 he officially withdrew as editor of the *Encyclopédie,* confining his future work in the project to completing the mathematical entries.[45]

During d'Alembert's long years of strife with rival grands géomètres and with hostile devots, salon society was his refuge from the storm, a setting where he felt perfectly at ease and was loved and admired by his companions. On occasion, however, the rivalry between the salons and their hostesses could become as bitter as the jealousies of great mathematicians, and it was d'Alembert's misfortune to be caught up in the midst of such a flare-up. In 1753 the aging Mme. Du Deffand was losing her eyesight, which made it difficult for her to perform her duties as a salonière. Not inclined to give up her position at the center of Parisian life, she decided to engage an assistant—her niece, Julie de Lespinasse, illegitimate daughter of Mme. Du Deffand's brother, the Comte de Vichy. The arrangement worked well in the early years; the two hostesses were close, and the salon continued to flourish. It remained the meeting place of the brightest stars in Parisian society, including Voltaire, Turgot, Rousseau, Montesquieu, and, of course, d'Alembert.

But with the passage of years the older hostess began to view her young relation not as an assistant but as a dangerous rival. Julie de Lespinasse, according to Condorcet, who knew her well, possessed a "true and simple sensibility," as well as "lively and natural social graces."[46] She was smart, charming and, of course, young, whereas Mme. Du Deffand was old and blind and in any case could never match the quick wit and grace of her niece. As a result, much to the older lady's chagrin, Lespinasse was gradually winning over the hearts of the most prominent salon guests. She forged an especially close bond with d'Alembert and became the most intimate friend he had in his lifetime.

The final rupture came in 1764 when Mme. Du Deffand discovered that the younger woman was entertaining the choicest guests in her own room a full hour before the start of the "official" salon gathering. This betrayal could not be tolerated, and Mme. Du Deffand demanded that her former protégé vacate her rooms in the rue St. Dominique immediately. Julie left, but not alone: she took with her many of the luminaries of the old salon and nearly the entire literary clique. In particular she took d'Alembert with her and set up a home with him in an apartment on the rue Bellechasse. There the two established their own salon, with Julie as hostess. It became known as the most socially informal, as well as the most philosophical, of Parisian literary salons.

Thus at the age of 47 d'Alembert finally left the home of Mme. Rousseau, the glazier's wife, and set up a home with his intimate friend, Mlle. Lespinasse. Condorcet, always eager to emphasize d'Alembert's natural goodness, tells us that although he moved out of the Rousseau household, d'Alembert never abandoned his simple foster mother who had nurtured him for so many years. "Twice every week," wrote Condorcet, "he returned to her to confirm with his own eyes the care she received in her old age . . . He sought to anticipate, to divine what would render the end of [her] life sweeter."[47] Meanwhile, in his new quarters his life changed less than one would expect. In place of Mme. Rousseau he now had Julie de Lespinasse to look after him, and their relationship, though intimate, was in all likelihood platonic. He continued as before, writing long hours during the day and socializing in salon society in the evenings.

D'Alembert and Mlle. Lespinasse lived together for 12 years, until her death in 1776 at age 44. Content in his bachelorhood and fully satisfied with their platonic friendship, he seemed to have believed that she was as happy with the arrangement as he was. Only after her death did he find the passionate love letters she had exchanged for years, first with the Spanish Marquis de Mora and later with the Comte de Guibert. Lonely and miserable, d'Alembert moved to an apartment in the Louvre to which he was entitled by virtue of his position as secretary of the Académie française When he died in 1783, despite the unhappiness of his final years, d'Alembert was a famous man: a leading mathematician, a cultural hero, and a friend and confidant of many of the most powerful men and women in France. The infant who had been abandoned on the steps of a church had become one of the leading personages of the realm.

Rousseau's Child

From his birth as an unwanted child abandoned to the mercies of strangers to his last days in the lonely apartment in the Louvre, the story of d'Alembert's life was a complex one. It saw great triumphs—mathematical, literary, and philosophical—that made him a national figure and a public face of the Enlightenment. It saw dazzling social success that raised him from an obscure academician to the toast of Parisian society and one of the best-connected and most influential people in France. But his life also saw petty jealousies and bitter disputes with his fellow mathematicians and vicious attacks from his political opponents. It also saw difficult and even heart-wrenching splits from old friends, including Diderot and Mme. Du Deffand, and the posthumous disillusionment with Julie de Lespinasse.

But in the telling of his life story by Condorcet, his friend and admirer, all such ambiguities are erased. In Condorcet's version d'Alembert is simply a natural man, pure of heart and unaffected by the rivalries and petty jealousies that inevitably grow in the fertile ground of society. In this story each and every stage in d'Alembert's remarkable rise from obscurity to greatness is made possible through his unaffected natural simplicity. For Condorcet, even the seemingly unfortunate circumstances of d'Alembert's birth were harbingers of his later greatness and were, in fact, what made it possible. His abandonment by his profligate mother, far from degrading him and turning him into an outcast, resulted in his being raised by the simple Mme. Rousseau as a member of her unassuming family. He grew up far from the immoral machinations of high society, absorbing instead the simple virtues of his working-class foster family. Seemingly without family or connections, and—officially, at least—"no one's child," d'Alembert was a child of nature herself, arriving in the world free and self-sufficient, unencumbered by the weight of social connections and obligations. As such, he was ideally suited to become a Rousseauian "natural man."

And so he proved; for d'Alembert, according to Condorcet, demonstrated his independence of social pressures and the purity of his heart at every turn in his life. Even while he was a mere student, d'Alembert followed his heart's calling to the study of mathematics, rebuffing all his teachers' efforts to steer him toward a life of the cloth. After his graduation he nevertheless made an effort to conform to social expectations by studying law and medicine, but to no avail—he could not help but follow his heart's desire and study mathematics. His choice to turn down lucrative careers in law and medicine demonstrated to Condorcet that d'Alembert would not, and perhaps could not, succumb to the pressures of social conventions. At the same time his dogged pursuit of mathematics even when, according to Condorcet, he believed that he lacked the talent to contribute to the field is also a case in point. It shows that far from pursuing glory and renown among his fellow men, d'Alembert was pursuing mathematics only because of his own deep-felt love of the subject. He was his own man, beholden to no one but himself, following the urgings of his own heart without regard to the expectations of others. "It is rare," Condorcet commented, "to observe the human heart so close to its natural purity, and before amour-propre corrupts it."[48]

D'Alembert's unconventional living arrangements, according to Condorcet, also showed him to be a simple "natural man." For over 40 years, even as he became a public figure and a leader of the philosophes, he made his home in the humble abode of Mme. Rousseau, the glazier's wife. "He lived there for nearly forty years," wrote Condorcet, "always preserving the same simplicity, not allowing the increase in his income to show except in his kind deeds . . . and always hiding his celebrity and glory so that his foster mother . . . never perceived that he was a great man."[49]

Finally, in Condorcet's view, d'Alembert's popularity in the salons and his rise to prominence were also due to his natural purity and simplicity. "Happy in the pleasure of his studies and his freedom, he preserved his natural gaiety in all its youthful innocence."[50] He was "content in his lot, desiring neither fortune nor distinctions," but his natural demeanor and unaffected gaiety made him an immediate success in *le monde* anyway. "They loved in him that simplicity [*bonhomie*] that can be found in superior men," but not in those who pretend to be so.[51] D'Alembert, according to Condorcet, shone in society precisely because he had no need of society. Because he desired nothing from his fellow men, he was immune to amour-propre, that all-pervasive ill of society that causes men to compare themselves with others and look beyond themselves for self-validation. Because he remained self-sufficient and true to himself, he proved irresistible to the overly cultured denizens of the Parisian salons.

For Condorcet, then, d'Alembert was the embodiment of a "natural man," and his life story demonstrates the remarkable effect such an individual can have on an overly civilized world. Because many of the anecdotes in Condorcet's "Éloge" came directly from his deceased friend, there is good reason to suppose that d'Alembert viewed his life in similar terms. His *Essai sur la société des gens des lettres et des grands,* in which he recommends that scholars follow a life of "liberty, truth, and poverty," also points to an outlook consistent with Condorcet's admiring account.

It shows that in theory, at least, d'Alembert considered self-sufficiency and disregard for social concerns necessary ingredients of a true scholar's life. Altogether it presents an ideal of a mathematician's life as that of a simple man who stands out among his peers for his unmediated connection to the world around him. Lesser men are lured by the siren song of society and become absorbed in its artificial and shallow world. But a mathematician like d'Alembert is not taken in by the temptations of civilized society and remains in touch with his true self and the real unadorned world around him. He is, in a word, a "natural man."

D'Alembert's advantage over his peers derived from the fact that he always remained a true child of nature, whereas they had been corrupted by civilized society. This advantage, furthermore, was the secret not only of his social success but also of his remarkable mathematical talents. For in his own eyes and those of Condorcet and his other admirers, d'Alembert's special mathematical understanding sprang directly from his unmediated relationship to the world around him. Mathematics, to them, revealed the hidden structures of the natural world, and no one was better suited to detect these than a pure "natural man." Surely it is no coincidence that d'Alembert's most important works were focused precisely on natural physical phenomena such as the theory of winds and the passage of water. D'Alembert's mathematics is "natural" because it springs directly from the natural world, just as he was a "natural man," remaining true to his roots in nature.

Natural Men

It must be remembered, however, that d'Alembert, prominent though he was, was only one of the grands géomètres of the eighteenth century. What of the others? Does the tale of the mathematician as a simple and happy "natural man," at home in the world and in society, apply to them as well? The question is important if we are to draw broader conclusions from d'Alembert's unusual biography and personality. If d'Alembert was the only leading mathematician of the age who was viewed by his contemporaries in those terms, then his unique narrative is representative only of his unusual biography and person. But if elements of his story can be found to apply to his mathematical peers as well, then we can perhaps begin to draw conclusions about the way in which mathematicians were viewed at the height of the Enlightenment and to compare it with the image of mathematicians at other times.

The remarkable story of d'Alembert's birth undoubtedly sets him apart from his peers, and no other grand géomètre of the age is associated with such a myth-laden story of origins. But in other respects d'Alembert's life and career are not as unusual among his mathematical peers and seem to embody certain prevailing trends in a purer and more distilled form. D'Alembert, as I have noted, became the shining star of the Paris salons, but he was not the first mathematician to enjoy social success in such settings. His friend and mentor Maupertuis became the toast of Parisian high society following his expedition to Lapland in 1735 and remained a popular and influential figure in these circles until he moved to Berlin in 1746. Condorcet, who was d'Alembert's protégé for many years until he embarked on his own career in public service, was also a familiar figure in Paris salons, which served as the source of his power and influence right up to the Revolution.

In the 1750s d'Alembert also became a public figure in his role as a leader of the philosophes in the French cultural wars. Although d'Alembert was more prominent in his public role than any of his mathematical contemporaries, nevertheless, both Maupertuis before him and Condorcet in later years also publicly championed the cause of the Enlightenment. Most other leading mathematicians, such as Clairaut, Daniel Bernoulli, and Lagrange, limited themselves to mathematical and scientific work, while Euler's rare ventures into broad philosophical discussions were generally not well received. Nevertheless, all saw themselves as bearers of a new spirit of enlightenment and rationality, and—despite differing degrees of religiosity—no leading mathematician was ever associated with the conservative opposition to the philosophes led by the French devots.[52]

As for the public personas of leading mathematicians, d'Alembert is again, to some extent, a man apart. None of his professional peers was ever described as a "natural man" in such explicit Rousseauian terms as d'Alembert was in Condorcet's "Éloge." Nevertheless, Condorcet's basic characterization of d'Alembert as a simple and cheerful man devoid of petty jealousies is quite typical of the public characterizations of Enlightenment mathematicians. Consider, for example, what Bernard Le Bovier de Fontenelle (1657–1757) had to say in his eulogy of Pierre Varignon, whose imprint on the Collège des quatre nations inspired d'Alembert's love of mathematics. As perpetual secretary of the Paris Academy of Sciences, Fontenelle held the same position in the early decades of the eighteenth century that Condorcet did in later years and was therefore responsible for eulogizing the academy's deceased members.

"His character was as simple as his superiority of spirit could ask," wrote Fontenelle, who continued: "He knew not jealousy . . . How many men of all sorts, elevated to the same rank, honored their inferiors by being jealous of them and decrying them! The passion to preserve the first place makes one take degrading precautions." But Varignon, according to Fontenelle, would never engage in such degrading behavior. Devoid of all pettiness, his behavior was "clear, frank, loyal on all occasions, free of all suspicion of indirect and hidden interest."[53]

Condorcet's 1783 eulogy of d'Alembert, as we have seen, was infused with flowery Rousseauian references to the man's purity and the corrupting influence of society. In comparison, Fontenelle's tribute to Varignon is much simpler, which is hardly surprising given that Rousseau, the prophet of natural sensibility, was only 10 years old when Varignon died. But in other respects the characterizations of Varignon and d'Alembert in their respective eulogies are unmistakably similar. Like d'Alembert, Varignon was incapable of selfish intrigue, was free of jealousy and pettiness, and always remained true to himself even when dissimulation might serve his interests better. Most significantly, like d'Alembert, Varignon is described as a "simple man," friendly and unaffecting, in touch with his fellows and his world.

The degree to which either Varignon or d'Alembert truly conformed to the idealized images presented by their friends can, of course, be questioned. As we have seen, d'Alembert's career hardly supports the notion that he was devoid of jealousy and ambition, and the description of Varignon as a simple, unassuming man might be equally exaggerated. This is to be expected, for eulogies by their very nature are intended to exalt, presenting the deceased in the best possible light, even at the expense of factual verity. They are deliberately tendentious texts, and the statements and descriptions in them should be treated with caution by anyone seeking accurate accounts of the people and events they describe. But as presentations of contemporary ideals of personality and conduct, eulogies can hardly be surpassed. They assume the prevailing notions of what an ideal man—or in our case an ideal mathematician—should be like and then do their best to fit the deceased into that mold.

What we learn from Fontenelle's eulogy of Varignon and Condorcet's eulogy of d'Alembert is not that both of these geometers actually possessed simple and noble souls; it is rather that in the eighteenth century the ideal geometer, or savant, possessed such a soul. By lauding their departed friends and colleagues, Fontenelle and Condorcet did only partial justice to Varignon's and d'Alembert's actual personality and conduct, but they did full justice to their own ideal of what a great geometer of the eighteenth century should be like: he was a simple and happy natural man who never succumbed to pettiness and jealousies but was deeply connected to his fellow man and his natural surroundings. One can hardly imagine a starker contrast with the brooding loners who populated the mathematical landscape of the following century.

Euler, as we have seen, was as different a man as can be imagined from his contemporary d'Alembert. Whereas d'Alembert was social and witty, Euler was awkward and plodding; whereas d'Alembert engaged in philosophy and literature, Euler limited himself almost exclusively to mathematical matters; while the Frenchman became a public figure and a leader in the cultural wars of the Enlightenment, his Swiss colleague limited his politics to (unsuccessful) attempts to secure the presidency of the Berlin Academy; finally, while d'Alembert, who hardly ever strayed from Paris, remained a bachelor all his life, Euler, despite his travels, found time to marry twice and father 13 children.

At the very end of their lives, however, the two grands géomètres did have a few things in common: they both died in the autumn of 1783, and both were the subjects of lengthy eulogies by the permanent secretary of the Paris Academy, Condorcet.[54] The two eulogies are very different in tone, which is not surprising given that Condorcet knew d'Alembert intimately but probably never met Euler. Understandably, the tribute to d'Alembert is far more personal and infused with lively anecdotes, whereas the tribute to Euler is rather formal and focuses on his mathematical achievements and his stature within the European community of savants. But given these contrasts in the careers and personalities of the two and in Condorcet's relationship with them, it is striking to see how similarly Condorcet characterizes them in his two eulogies.

Euler, according to Condorcet, always exhibited a "simplicity" and an "indifference to renown."[55] "Never in his learned discussions with celebrated geometers," Condorcet added, "did he let escape a single act that would make one suppose that he was occupied with the interests of his amour-propre."[56] He then went on to describe Euler's remarkable generosity to his colleagues and his willingness always to acknowledge his rivals' claims of priority. This, of course, is also precisely what Condorcet said about d'Alembert, the "natural man" who was free of petty jealousies and devoid of amour-propre. Finally, Condorcet wrote, Euler embodied the rare union of unblemished happiness with uncontested glory.[57] This again is very reminiscent of Condorcet's account of d'Alembert, whose purity of heart manifested itself in an unaffected gaiety of spirit.

In their eulogies, at least, Varignon, d'Alembert, and Euler were all simple, happy men who fully engaged with their fellows and lived and enjoyed life to the fullest. They were all "down to earth" in the best sense of the term—well grounded and connected to their surroundings, both natural and human. Because this uniformly attractive picture is applied to such different men, with radically different personalities and careers, we can conclude that this description does not apply to a particular person but is rather an ideal type. It does not tell us who a particular mathematician was, not even if we make allowances for the laudatory conventions that govern the writing of eulogies. It tells us, rather, what an ideal mathematician should be in the eyes of his contemporaries: a simple, modest man, a man of the earth and a child of nature.

Certainly, not everyone agreed with this idealized vision of a geometer as a man deeply in touch with his surroundings. Diderot, who grew increasingly disenchanted with mathematics during his collaboration with d'Alembert on the *Encyclopédie,* thought that the grands géomètres were so absorbed in their rarefied world that they had completely lost sight of reality. They "resemble a man who watches from those mountain tops whose summits are lost in the clouds," Diderot wrote in 1753. "The objects visible on the plain have disappeared before his eyes, so that all that remains with him is the spectacle of his own thoughts."[58] Sixteen years later, when he wrote *La rêve de d'Alembert,* Diderot was equally dismissive of mathematicians and the role they could play in the discovery of new knowledge. D'Alembert, in his account, is a confused bumbler who barely knows where he himself is and needs the constant assistance of the grounded Mlle. Lespinasse to survive. For Diderot, the notion that such a man possesses a deep understanding of reality is ludicrous, and d'Alembert's ideas of the natural world are accordingly overly abstract and spurious.[59]

Diderot's view of mathematicians was taken on in the following century in the popular myths of young luminaries such as Évariste Galois and Niels Henrik Abel. Like the bumbling d'Alembert in Diderot's pamphlet, these tragic mathematicians were also fully absorbed in their imaginary mathematical world and were, as a result, disconnected from their earthly surroundings. The value judgment placed on this disconnected state is, however, radically different in the two cases. In the nineteenth century, as we shall see, the fact that brilliant mathematicians did not fit well into their mundane social setting was viewed as a sign of transcendent genius. But for Diderot in the eighteenth century, d'Alembert's overabsorption in the rarefied world of mathematics simply made him irrelevant to the progress of knowledge. D'Alembert, in Diderot's account, is a figure of fun, lost in a cloud of irrelevant formulas.

Like Condorcet's eulogy, Diderot's parody cannot be taken at face value as a description of d'Alembert. Whereas Condorcet's account is a calculated aggrandizement of the man, Diderot's version is designed to amuse and belittle. Nevertheless, both Condorcet and Diderot knew d'Alembert intimately, and if the eulogy still carries residues of the living man, it is likely that the parody does as well. Was d'Alembert then truly a "natural man" closely engaged with his surroundings, as Condorcet claims? Or was he, as Diderot suggests, a disconnected dreamer and a hopeless bumbler in the company of men? We will never know, although it is reasonable to assume that he was a bit of both. His meteoric rise in society and his prominence among the philosophes seem to support Condorcet's view; the fact that he lived in his foster mother's home well into his forties and his astonishment at his discovery of Julie de Lespinasse's love affairs suggest the truth of Diderot's view.

D'Alembert's "true" persona (if such a thing there is) will remain forever beyond our reach. For our purposes, however, it hardly matters, for both Condorcet and Diderot were using d'Alembert's life and personality not just to preserve a record of the man himself but also to present their vision of a paradigmatic mathematician. When viewed in this light, it appears that the two faces of d'Alembert reflect two opposing visions of mathematics that competed for ascendancy in the Enlightenment. On the one side were those, like d'Alembert himself, but also Voltaire, Maupertuis, and, later, Condorcet, who believed that mathematics and its strict form of reasoning held the key to knowledge of the world; on the other side were Diderot and the Comte de Buffon, who believed that mathematics dealt with abstractions that were largely irrelevant for the study of the world as it truly is.

Significantly, the opposing sides were in agreement on the fundamental standard by which mathematics should be judged: its relevance to the understanding of the world. The admirers of the field claimed that it was relevant to the world, whereas its detractors argued that it was not. Neither side considered the position that became popular in the nineteenth century, that mathematics must be judged by its own pure and rigorous standards, and that its usefulness to the study of the world is irrelevant to its truth value.

Naturally, d'Alembert and the other leading mathematicians of his age were all promoters of the value and importance of their field. In the eighteenth century this meant an insistence that mathematics was profoundly relevant to the study of the world, and that mathematicians possessed a unique insight into its hidden workings. Varignon, Euler, and d'Alembert were therefore idealized as men who were deeply connected to nature and to their surroundings. To their admirers, they were grounded "men of the world," comfortably at home in both the natural and human domains. As we shall see in the following chapter, this outlook also helped shape the actual mathematics practiced by Enlightenment geometers, the kinds of problems they took on, and the kind of solutions they considered relevant.

Just as significantly, the ideal of the practical and natural man also shaped how the grands géomètres saw themselves, and they did their best to live up to this image. Indeed, one must admit that on the whole, the leading Enlightenment mathematicians were a practical lot, well integrated into society and well capable of taking care of their interests. A tiny clique in themselves, they managed to obtain and keep the most prestigious chairs at the most prestigious academies and centers of learning in Europe. The Bernoulli clan is perhaps the most outstanding example of this success, having managed to install its members in prestigious academic positions for four generations, but the personal success of others was at least as impressive: Varignon was a member of the Paris Academy of Sciences, as well as the Berlin Academy and the Royal Society of London; Maupertuis rose to become the president of the Berlin Academy, and d'Alembert was courted to replace him by none other than Frederick II of Prussia; Euler was a member of all the great academies of Europe and held chairs in St. Petersburg and Berlin; and Lagrange moved from the Academy of Turin to Berlin and then to Paris before becoming a high official in the Revolution. Quite as significantly, these leading mathematicians were close acquaintances, and sometimes even friends, of the great rulers of the age. Both Maupertuis and d'Alembert, for example, were on intimate terms with Frederick II of Prussia, and Euler had befriended

Catherine the Great of Russia. In their self-image and in practice, the grands géomètres were a practical and successful lot, true to their self-image as natural men.

But not one of the grands géomètres could match the myth and the charismatic aura that surrounded d'Alembert. As a parentless child, a foundling raised by a foster family and seemingly without family or connections, he fashioned himself as a man unbound by man-made conventions and therefore more "natural" than his fellows. His humble living arrangements, his social success with salon hostesses and monarchs alike, his dogged pursuit of the field he loved, and even his exceptional mathematical talents all testified that he was truly a "natural man" untouched by amour-propre. To his admirers, d'Alembert was always a child, retaining an unmediated connection to the world, both human and natural, that is lost to the rest of us. And whereas most men erected barriers of social conventions between them and nature, d'Alembert possessed an unmediated connection with the natural world and a deep understanding of its deep structures. Such was the image and the myth of Jean le Rond d'Alembert, Enlightenment geometer and Rousseauian natural man.

2

NATURAL MATHEMATICS

The View from the Mountain

What kind of mathematics was practiced by the "natural men" of the Enlightenment? It was, in brief, a "natural mathematics," a science of the world that had its roots planted firmly in material reality. For d'Alembert and his contemporaries, all mathematical relations, abstract though they were, were ultimately derived from the physical relations existing in the world around us and perceived by our senses. Conversely, they believed, general mathematical relations, derived at the highest levels of mathematical abstraction, informed us about the true realities of our physical world. For eighteenth-century practitioners, mathematics was derived from the world, while at the same time the material world itself was structured according to mathematical principles. This dual relationship between mathematical abstraction and the natural world was the defining characteristic of Enlightenment mathematics.

It is less than surprising to find that the best account of the meaning and character of Enlightenment mathematics was provided by d'Alembert, the paradigmatic philosophe and mathematician of the age, and the field's most celebrated spokesman. His defense of mathematics was published in 1751 as part of the "Preliminary Discourse" to the *Encyclopédie,* which secured his status as a leading philosophe and led to his election to the Académie francaise. According to d'Alembert, mathematics is simply a description of physical objects when these are considered in the abstract, that is, without some of their material attributes. Geometry, for example, is the science of objects considered as pure extension, shorn of their quality of "impenetrability." Arithmetic is born of the necessity of comparing several geometrical objects: "It is simply the art of finding a short way of expressing a unique relationship [a number] which results from the comparison of several others." Algebra, in turn, is an abstraction of arithmetic, expressing its relationships in a universal general form.[1] This, according to d'Alembert, is where the chain of abstractions stops: "This science [algebra] is the farthest outpost to which contemplation of the properties of matter can lead us, and we would not be able to go further without leaving the material universe altogether."[2] Abandoning the physical moorings of mathematics, however, was evidently unthinkable for d'Alembert. It would empty the field of its content and meaning.

Once we have arrived at this peak of abstraction, d'Alembert continued, we can turn back and retrace our path. Step by step we now restore to mathematics the physical attributes that we had previously removed. Thus, after investigating the properties of geometrical objects in the abstract, we restore to them impenetrability, "the last sensible quality of which we had divested it": "The restoration of impenetrability brings with it the consideration of the actions of bodies on one another, for bodies act only insofar as they are impenetrable. It is thence that the laws of equilibrium and movement, which are the object of Mechanics, are deduced."[3] The physical sciences, then, for d'Alembert, are simply abstract mathematics, once the physical attributes, such as extension, impenetrability, and motion, are restored to its objects.

Despite d'Alembert's confident statement, however, the question of the relationship of mathematical abstraction to physical reality remained a troubling one to Enlightenment mathematicians. Since higher mathematics appeared set on a road toward ever-increasing abstraction, would it be able to find its way back to the physical world? Would the restoration of physical qualities to abstract mathematics actually result in a true science of nature? D'Alembert certainly believed this to be the case, but even he was obliged to admit that this was not self-evident, and this concern is very much in evidence in the "Preliminary Discourse." After surveying the landscape of human knowledge, he asks the reader to "stop and glance over the journey we have just made." Looking back, he writes, one will notice two "limits," or landmarks, of absolute certainty. One of these is "our idea of ourselves," the Cartesian "I."[4] The other, says d'Alembert, is abstract mathematics, "whose object is the general properties of bodies, of extension and magnitude." He continues:

> Between these two boundaries is an immense gap where the Supreme Intelligence seems to have tried to tantalize the human

> curiosity, as much by the innumerable clouds it has spread there as by the rays of light that seem to break out at intervals to attract us . . . We are indeed fortunate if we do not lose the true route when we enter this labyrinth! Otherwise the flashes of light which should direct us along the way would often serve only to lead us further from it.[5]

Even for the optimistic d'Alembert, the landscape of knowledge presents a formidable challenge to the inquisitive philosopher. Only two peaks of certainty rise above the obscuring clouds, offering secure vantage points. Surveying the vast undiscovered country below, one observes only small patches of clarity, surrounded by vast stretches of misty and confusing terrain. The mission of the natural philosopher is to turn his gaze from the secure "limit" of his mathematical knowledge toward this "immense gap" and very carefully seek a true path through the labyrinth.

For d'Alembert, then, mathematics is a secure vantage point of clarity and reason from which to look down on the misty valley of uncertainty below. The penetrating force of the "mathematical vision" will break through the clouds of confusion and bring the mysteries of the natural world under the enlightened scrutiny of mathematical reason. From the lofty heights of abstract mathematics one can look down onto the physical world and unveil the hidden mathematical patterns that structured it. Since d'Alembert and most of his fellow geometers believed that the world was essentially mathematical, the ultimate success of the venture was assured.

Not all philosophes shared d'Alembert's confidence that mathematics is the key to knowledge of the physical world. In 1749 the leading naturalist, the Comte de Buffon, argued that mathematicians are trapped in their own bubble of abstraction, doomed forever to repeat their initial assumptions in ever more sophisticated ways.[6] Closer to home for d'Alembert, there is little doubt that Diderot, his friend and rival, had the "Preliminary Discourse" in mind when he wrote in 1753 that mathematicians "resemble a man who watches from those mountain tops whose summits are lost in the clouds; the objects visible on the plain have disappeared before his eyes." In saying this, Diderot was not disputing d'Alembert's main contention, that mathematicians are those who travel to the very limits of the world of knowledge and enjoy pristine clarity at the heights of mathematical abstraction. His criticism was precisely on the point that seemed to concern d'Alembert as well: whether the mathematician will ever find his way back from the heights of abstraction down to the "real world" to throw some light on its hidden secrets. D'Alembert thought that this was possible (though there are indications that late in his career he had some doubts); Diderot thought that abstract mathematics was so far removed from the physical world as to be practically irrelevant to our knowledge of it.[7]

Diderot's criticism of the "mathematical vision" went so far that he likened mathematicians literally to those born without eyesight. In his "Letter on the Blind" of 1749 he argued that the ideal mathematician was Nicholas Saunderson (1682–1739), who despite being blind from infancy rose to become Lucasian Professor of Mathematics at Cambridge. Saunderson, Diderot pointed out, had an amazing capacity for abstract thinking that could hardly be matched by his seeing colleagues. His isolation from the outside world did not hinder but rather helped Saunderson in his abstract mathematical reasoning. But at the same time, Diderot argued, this insularity made his work irrelevant to our understanding of physical reality. For d'Alembert, a mathematician was a sharp-eyed observer on a mountaintop who could see further and deeper than others; for Diderot, he was a blind man, trapped within the confines of his own mind.[8]

By insisting in the "Preliminary Discourse" that mathematics is natural, rather than artificial, and fundamentally worldly, d'Alembert was taking a stand for his field. Faced with critiques from fellow philosophes that mathematics is too abstract and insular to contribute to our knowledge of the world, d'Alembert insisted on the centrality of the field to the new enlightened sciences. For him, the heights of mathematics provide a grand panoramic view of the physical world, revealing the symmetries and harmonies that lie hidden within it. Far from being a disembodied set of abstract relations, as Diderot and Buffon had claimed, mathematics is for d'Alembert an inseparable part of the physical universe.

D'Alembert and his fellow geometers fashioned themselves as the unspoiled children of nature of Rousseauian myth. Immune to the corruption of society and the destructive effects of amour-propre, they were childlike innocents who retained a child's unmediated access to the world around him. Whereas most men view the world through the veils of social corruption, Enlightenment mathematicians presented themselves as men who retained a childlike access to the natural world and were uniquely attuned to its rhythms and deep structure. They were "natural men," and inevitably their practice was "natural mathematics": the study of the deep harmonies that govern the world, invisible to most men, but accessible to natural men like themselves.[9]

The March of Abstraction

With the exception of d'Alembert, most of the grands géomètres of the Enlightenment never bothered to write down their philosophical views on their field. Like mathematicians in all ages, they preferred to practice mathematics rather than spend their time writing philosophical tracts. Their work, however, speaks for itself, making it clear that from Varignon to Lagrange, eighteenth-century geometers fully shared d'Alembert's view of the nature and purpose of mathematics. The movement of mathematics throughout the eighteenth century was from the geometrical and concrete to the algebraic and abstract. But even in its most abstract forms practiced by Euler and Lagrange, Enlightenment mathematics remained essentially about the world as it is. Physical reality was both the object of mathematical studies and a guarantee of the truth of its claims: if a mathematical system correctly described the world, then it must be mathematically true as well.[10]

Early in the century, when the calculus introduced by Newton and Leibniz was still a novelty, mathematicians such as the Marquis de L'Hôpital (1661–1704), Pierre Varignon, and Jakob Bernoulli (1654–1705) focused on questions that were essentially geometrical. What, they asked, is the precise curve described by a chain hanging from two points? What is the shape of the brachistochrone—the path of a body moving from one point to another on a different plane in the least amount of time, under the force of gravity? Or of the tractrix—the path of a body dragged by a tight string over a resisting medium when the string is pulled at a right angle to the original line between the body and the puller? Or of the cross section of a sail inflated by the wind?[11]

Later in the century the questions addressed by Continental Enlightenment mathematicians took on a more general character.[12] Instead of addressing specific geometrical and mechanical questions, as their predecessors had done, midcentury mathematicians sought to define mathematically the laws governing whole branches of mechanics. Using the new and powerful tool of differential equations, d'Alembert and his generation introduced entire new fields of investigation, best described as "theoretical mechanics." Euler's *Mechanica* of 1736, which presented

Newtonian mechanics in terms of differential equations, was an early example of this trend, and it was followed two years later by Daniel Bernoulli's *Hydrodynamica,* the first attempt to describe the flow of liquids using the tools of analysis. D'Alembert in his *Traité de dynamique* (1743) sought to do the same for dynamics, and Clairaut's *Théorie de la figure de la terre* (1743) mathematically derived the shape of the earth from Newtonian principles.

As the questions asked by Enlightenment geometers became more general, their methods also became more abstract. Early in the century both the questions and the solutions were fundamentally geometrical in nature, and mathematicians considered the calculus a method for investigating specific curves, their slopes, and their enclosed areas. After 1730, however, the emphasis moved away from geometry and increasingly toward algebraic methods. Clairaut, Euler, and d'Alembert were pioneers of the field of algebraic analysis, which treated the calculus in terms of algebraic expressions and the relationships between them. A given formula, or *function,* as such algebraic expressions came to be called, was related to other functions through the inverse operations of differentiation and integration. The "calculus," which was tied to specific geometrical constructions, now became "analysis"—a general algebraic method that could be applied to many different problems. For d'Alembert and his contemporaries, the role of the analyst was to make use of this powerful algebraic method to describe the structure and operations of the physical world. In doing so, mathematicians were revealing the hidden mathematical order that they believed structured the world around us.

By the end of the century analysis had moved even further from its geometric roots. First Euler and then Lagrange insisted that analysis should be liberated from any geometrical representation whatsoever and become purely algebraic. "Mathematics, in general, is the science of quantity; or the science that investigates the means of measuring quantity," Euler wrote in 1771.[13] Algebraic relations, he believed, describe relations between any set of quantities whatsoever and are therefore applicable to any problem that involved quantities. The power of algebra, and of algebraic analysis in particular, lay in its complete generality, which made it an incomparable tool for investigating the myriad phenomena of the physical world. Whereas geometrical arguments were necessarily limited to particular cases and instances, algebraic analysis was by its very nature general and universally applicable.

It is for that reason that Lagrange, who brought the Enlightenment tradition of analysis to its highest perfection, proudly wrote in the introduction to his classic *Mécanique analytique* of 1788:

> No figures will be found in this work. The methods I present require neither constructions nor geometrical or mechanical arguments, but solely algebraic operations subject to a regular and uniform procedure. Those who appreciate mathematical analysis will see with pleasure mechanics becoming a new branch of it and hence, will recognize that I have enlarged its domain.[14]

It is precisely the removal of all visual representations, Lagrange argued, that makes algebraic analysis so powerful. Devoid of specific geometrical representations, analysis is free to restate the laws of mechanics in its own general and abstract terms, making mechanics, in effect, a branch of analysis.[15]

Overall, then, the trajectory of mathematical development in the eighteenth century resembled d'Alembert's structural account in the "Preliminary Discourse." Geometry, arithmetic, and algebra, d'Alembert wrote, simply represented increasing levels of abstraction from the physical world, with algebra being "the farthest outpost to which contemplation of the properties of matter can lead us . . . without leaving the material universe altogether."[16] The actual historical development of eighteenth-century mathematics closely followed his account, moving along a progression of increasing abstraction. Beginning with a geometrical emphasis early in the century, it had clearly arrived at the "farthest outpost" of abstraction when Lagrange published his *Mécanique analytique*. That, however, was as far as Enlightenment mathematicians would go; leaving the material universe altogether was to them unthinkable.

The Catenary

For a taste of how mathematics was practiced around 1700, consider the problem of the catenary—the shape of a chain hanging from two points. Galileo had considered the question in 1638 and had proposed that the shape of the catenary was in fact a parabola, and that a hanging chain could therefore be used to draw a parabola quickly and accurately.[17] Galileo had offered no demonstration of this assertion, and by the 1660s other mathematicians had demonstrated that the curves could not be one and

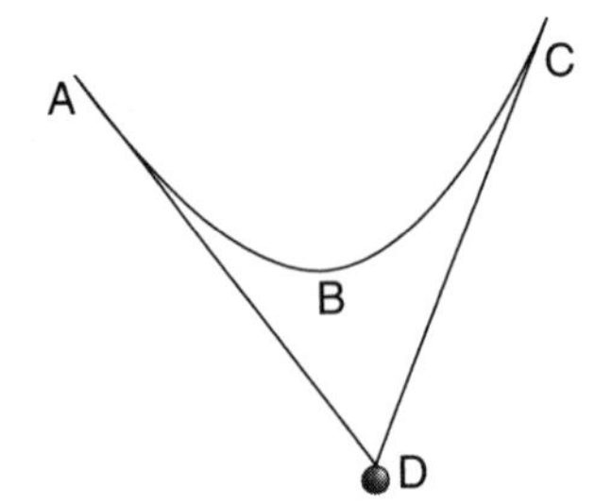

Figure 2.1. Johann Bernoulli's weighted string reproducing the forces at points *A* and *C* on the catenary. Johann Bernoulli, *Johannis Bernoulli opera omnia,* vol. 3 (Lausanne: Bousquet, 1742), 498.

the same. The true character of the catenary, however, remained unknown until Jakob Bernoulli posed it as a public challenge to mathematicians in 1690. Among those to respond were the greatest geometers of the age—the Dutch polymath Christiaan Huygens (1629–1695), Gottfried Wilhelm Leibniz (1646–1716), and Jakob's brother Johann (1667–1748).[18]

Johann Bernoulli's strategy for determining the shape of the catenary was, first, to come up with an equation describing the tangent of the curve at every point. According to the practice of the early Leibnizian calculus, the way to accomplish this was to add an infinitesimal increment, or *differential,* to the curve at a random point. The corresponding increments along the x and y axes were known as differentials in Leibniz's calculus and designated dx and dy, respectively. The ratio between them, $dy:dx$, is the slope of the tangent at that given point, and a general equation determining $dy:dx$ in terms of the coordinates x and y determines the tangent of the curve at every point along it. From this expression, by a process of integration, the curve that possesses just such a tangent at every point over its entire length can be found, and that is the catenary.[19]

Characteristically for this era, Johann Bernoulli treated the mathematical problem of the catenary as a mechanical one: the curvature of the chain, he reasoned, is shaped by the force exerted by the hanging chain at each and every point along its curve. Accordingly, he posited a chain hanging from two arbitrary points, *A* and *C,* with its lowest point of the curve designated *B* (Figure 2.1). The forces exerted by the chain on these two points, he argued, are equivalent to the forces exerted on them by a single weight equivalent to the weight of the chain, hanging by a weightless string from *A* and *C* and stretching precisely along the curve's tangents at these points (Figure 2.1). This is obvious, Bernoulli argued, because the chain itself also pulls on the points *A* and *C* with the exact same weight, at precisely the same direction—that of the tangents.

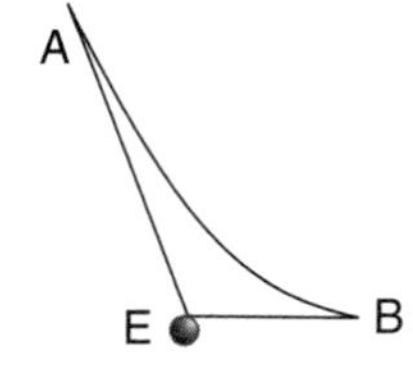

Figure 2.2. The weight of the chain *AB* hanging by weightless strings at the intersection *E* of the tangents *AE* and *EB*. Johann Bernoulli, *Johannis Bernoulli opera omnia,* vol. 3 (Lausanne: Bousquet, 1742), 498.

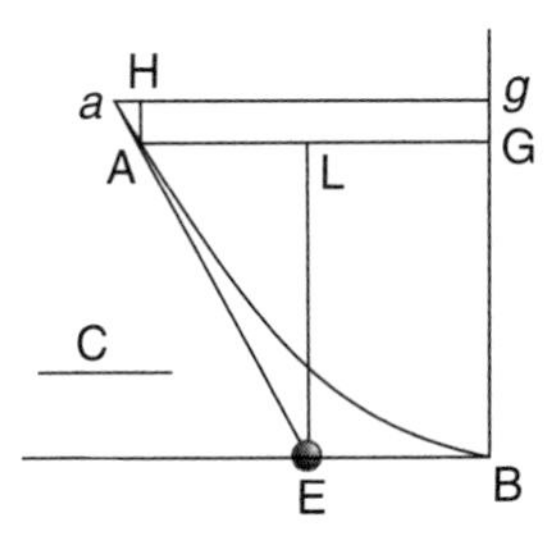

Figure 2.3. The balance of forces on the catenary *AB*. Johann Bernoulli, *Johannis Bernoulli opera omnia,* vol. 3 (Lausanne: Bousquet, 1742), 498.

Bernoulli then repeated the process, this time looking at the span of chain *AB,* where *B* is the lowest point of the span *ABC*. As in the previous case, he posited a weight equal to that of the chain length *AB* at the intersection *E* of the tangents from *A* and *B* (Figure 2.2). The chain is once more held in place by weightless strings connecting it to *A* and *B,* respectively, the tangent *EB* this time being horizontal. Once again, the forces exerted at A and B by the weighted string are equivalent to the forces exerted on them by the chain.

At this point Bernoulli turned to the new methods of the infinitesimal calculus. Looking at point *A,* he extended the curve there by an infinitesimal increment *Aa* (Figure 2.3). The component of this increment along the *x* axis is *AH,* which is accordingly designated the differential *dx*. The component along the *y* axis is the segment *Ha,* which is accordingly designated the differential *dy*.[20] The slope of the catenary at A is therefore the ratio $dx{:}dy = AH{:}Ha$.

He then looked at the forces operating at the point *E,* balanced between the vertical pull of the chain's weight and the horizontal pull that the tight string is exerting at *B*. According to the parallelogram of forces, the ratio of forces at *E,* (weight at *E*):(the force exerted on *B*), equals the ratio *EL*:*AL,* which itself is equal to the ratio of the increments along the *x* and *y* axes, *AH*:*Ha*. In other words, the ratio of the forces acting at *E* is precisely the same as the ratio of forces acting at *A* and determining the slope of the curve at that point. In mathematical terms this means that

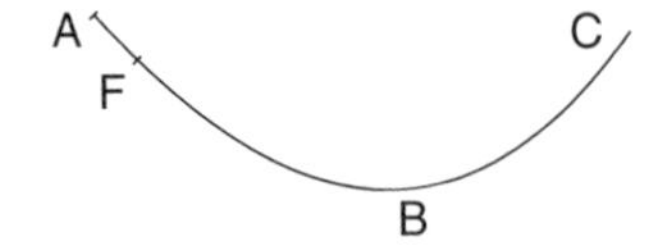

Figure 2.4. The catenary remains unchanged whether it hangs from *A* or from *F*. Johann Bernoulli, *Johannis Bernoulli opera omnia*, vol. 3 (Lausanne: Bousquet, 1742), 498.

$$dx : dy = (\text{weight at } E):(\text{the force exerted on } B). \tag{2.1}$$

Now, Bernoulli pointed out, the force exerted at *B* is always the same, no matter which arbitrary point *A* along the curve is chosen for drawing the tangent, and at which point *E* that tangent crosses the horizontal tangent. We know this, Bernoulli argued, because the shape of a catenary does not change, regardless of whether we lengthen or shorten the chain at the top. If, for example, we chose a point *F* along the catenary, part of the way between *A* and *B*, fixed it in place, and hung the chain from it instead of from *A*, the curve between *F* and *B* would not be affected at all (Figure 2.4). This fact, Bernoulli argued, "requires no demonstration; for reason persuades us of it, and the daily experience of the eyes establishes it."[21]

From this feature Bernoulli concluded that the forces acting at every point on the catenary are fixed, no matter what the total length of the chain is. At the bottom of the chain, at *B*, the hanging weight of the chain is zero, and the only force operating is along the horizontal tangent *BE*. Bernoulli therefore designated the tension along *BE* by the constant *a*, which gives

$$dx : dy = (\text{weight at } E) : a. \tag{2.2}$$

If we now further consider that the weight of a homogeneous chain is proportional to its length *AB*, and designate this length *s*, we arrive at

$$dx : dy = s : a. \tag{2.3}$$

Bernoulli then noted that the infinitesimal right triangle *Aha* gives $ds^2 = dx^2 + dy^2$, where *ds* represents *Aa*—the infinitesimal increment in the length of the chain at *A*. From this, using methods of substitution, Bernoulli eliminated the variable *s* in Equation 2.3 and derived the differential equation

$$dy = \frac{adx}{\sqrt{x^2 - a^2}} \tag{2.4}$$

that describes the slope $dx : dy$ of the catenary at every point.

To determine the shape of the catenary itself, Bernoulli still needed to integrate this equation and find the algebraic expression for which this slope holds true. Doing so was no mean feat using the methods of the 1690s. Modern mathematicians, using methods of algebraic integration developed later in the eighteenth century, easily arrive at the algebraic expression of the catenary,

$$x = \frac{a}{2}\left(e^{\frac{y}{a}} + e^{\frac{-y}{a}}\right). \tag{2.5}$$

But Johann Bernoulli did not have the advantage of these later developments. Instead of algebra he relied on tried-and-true geometry, transforming $dy = adx/\sqrt{x^2 - a^2}$ into an expression that contains the equation of an equilateral hyperbola with a as a parameter. He then showed that the area under this hyperbola, taken at each point x, provides the corresponding value of y for the catenary. By this means Bernoulli transformed the problem of determining the shape of the catenary into the determination of the area under a hyperbola—or the "squaring" of a hyperbola, as it was called at the time.

So ends Johann Bernoulli's resolution of the problem of the catenary. And a remarkable resolution it is: here algebraic equations live side by side with physical weights, geometrical constructions with mechanical forces and tight strings. The geometrical construction of the catenary is for Bernoulli an abstraction from the physical reality of a hanging chain, and the algebraic expressions are a shorthand for the geometrical constructions. For Bernoulli, furthermore, all these levels are so closely integrated with each other that they are practically interchangeable. The physical hanging chain is abstracted into a geometrical curve, but the slope of this geometrical curve is then determined by physical forces acting on material strings. The geometrical curve of the catenary is generalized into an algebraic differential equation, but this equation is ultimately solved not algebraically, but through a geometrical construction. And when general mathematical arguments are in short supply, Bernoulli does not hesitate to call on "daily experience" and the evidence of the eyes to buttress his argument.

Bernoulli moves easily back and forth between these layers of abstraction, never doubting for a moment that they perfectly correspond to each other. If a relationship exists in the physical world, it will also be true in a geometrical construction or a mathematical equation. Conversely, an algebraic operation such as differentiation or integration will necessarily correspond to geometrical constructions and mechanical forces. This has to be the case, Bernoulli believed, for what is mathematics but a generalization from the physical world? This was the guiding vision of the very worldly mathematics of Johann Bernoulli and his contemporaries. It was also the view that d'Alembert presented decades later, in elegant and persuasive prose, in the "Preliminary Discourse."

Polygons, Curves, and Vibrating Strings

The underlying assumptions of eighteenth-century mathematics remained largely unchanged throughout the century. Detractors of the field, such as Buffon and Diderot, claimed that mathematics was a self-contained field, occupied with expressing its own assumptions in ever more elaborate and convoluted ways. As such, they argued, mathematics was detached from reality and therefore irrelevant. In response, the promoters and practitioners of mathematics asserted that far from being a self-contained field, it was anchored in the material world, and its relations expressed actual physical realities. D'Alembert was, as we have seen, the most eloquent and celebrated champion of this view, but its essentials were shared by all the leading mathematicians of the time. Most significantly, as we have seen in the case of Johann Bernoulli's treatment of the catenary, the view guided the actual practice of mathematics in the Enlightenment.

The close interrelationship between mathematical concepts and their physical counterparts was also evident in the debate over the proper way to determine the tangent to a curve.[22] The trouble arose from the manner in which the Leibnizian calculus treated curves, and it greatly vexed many eighteenth-century mathematicians, including the Marquis de L'Hôpital, Pierre Varignon, Jakob Hermann (1678–1733), and d'Alembert. According to the practice of the early calculus, the way to determine the properties of a given curve is to break it up into straight lines, creating a "polygon curve" approximating the original. The greater the number of lines making up the polygon, the closer the fit between the new curve and the smooth original. Now according to Leibniz, the way to determine the tangent of the curve at a point is to continue the line of the polygonal curve that touches at that point. As the number of sides of the polygon is increased, the polygonal tangent moves closer to the true tangent at that point, and when the number of sides is infinite, the polygonal tangent is equal to the true tangent.[23]

This would have been perfectly satisfactory had the question been viewed as purely mathematical. But for eighteenth-century geometers, mathematical constructions had meaning in the real world, and the tan-

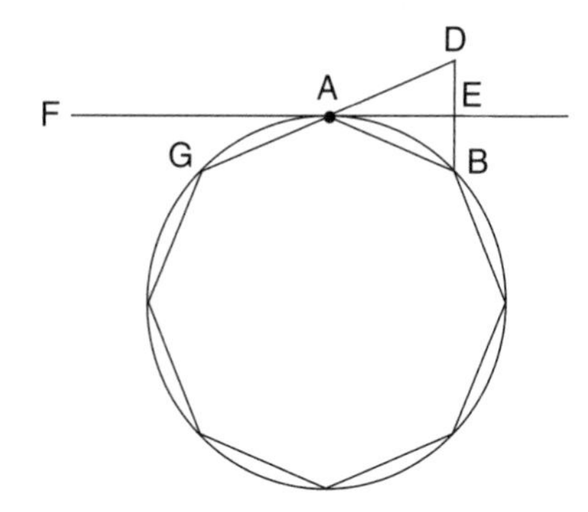

Figure 2.5. The polygonal curve and the tangent of a circle. After Thomas L. Hankins, *Science and the Enlightenment* (Cambridge: Cambridge University Press, 1985), 22, figure 2.2.

gent of a curve had a specific and well-known interpretation: it was the trajectory of a body traveling along the curve if it was released at the point of the tangent. As Newton had demonstrated in the *Principia mathematica* of 1687, the centripetal force required to keep the body traveling along the curve is proportional to the distance between the curve and its tangent. But here's the problem: the distance of the polygonal tangent from its curve is always double the distance of the true tangent from the original curve, no matter how many sides the polygon has or how closely it approximates the true curve. To eighteenth-century mathematicians, this posed a serious dilemma: Is the centripetal force acting on a body proportional to the distance of a true curve from its tangent? Or is it proportional to the distance between the polygonal curve and the tangent, and therefore double the value? In Figure 2.5 the distance *DB* between the polygonal curve and its tangent *GD* at *A* is double the distance *EB* between the curve and its true tangent at *A, FE*. This is true no matter how many sides are added to the polygon and how closely it approximates the true curve.

The source of the problem lay in the tacit assumption that mathematical relations corresponded to physical reality. If mathematical constructs had been accountable only to mathematics, then the problem would have disappeared: the polygonal method would yield results that were consistent with its own assumptions, whether these accorded with a certain physical interpretation or not. But for eighteenth-century practitioners, mathematical abstractions were inseparable from their physical manifestations and had to be consistent with them. Because the centripetal force acting on a body moving along a curve was always the same, the mathematical operations corresponding to this reality must yield a correct and consistent result.

Vexed with the conflicting mathematical results, at least one geometer sought to solve the problem through physical methods. In 1732 Jacques Eugène d'Allonville, Chevalier de Louville (1671–1732), a respected academician and correspondent of Isaac Newton, proposed an experiment to the Paris Academy of Sciences to settle the issue once and for all. All that was required, he argued, was to observe whether a ball moving in a circular orbit and then released would fly off along the tangent or in a different direction determined by the polygonal curve. The idea was never taken up by the academy, but in the eighteenth century it was not as far-fetched as it might appear to us today. After all, if mathematical operations were inseparable from material reality, and if physical considerations such as weight and force played a role in mathematical reasoning, then why should not physical experimentation indicate correct mathematical results? Louville was merely pushing prevailing views about mathematics to their logical conclusion.[24]

Louville's attitude was echoed some years later by Daniel Bernoulli, a second-generation member of the Bernoulli mathematical clan and son of Johann, who had resolved the catenary problem in 1692. The year was 1753, and Bernoulli was entering the fray of the debate over vibrating strings that had split the community of mathematicians in Europe since d'Alembert had published his groundbreaking paper on the subject in 1746. D'Alembert had provided the basic differential equations describing the motion of a vibrating string, and no one in the subsequent debate challenged the correctness of his solution. But in further exchanges on the topic a fundamental argument emerged between d'Alembert, on the one hand, and Euler, on the other, over the proper starting conditions of the hypothetical string. D'Alembert insisted that in order to be mathematically treatable, the string must conform to strict parameters, which in modern terminology would make it continuous and everywhere differentiable. Euler responded that the curve of the string could be any function whatsoever, even one "drawn by hand." The debate raged for years, with most of the grands géomètres of Europe lining up on one side or the other of the issue.[25]

From a modern perspective it appears that the debate had little to do with the physical question of the vibrating string and everything to do with the purely mathematical issue of the nature of a function, but that is not how contemporaries saw it. For eighteenth-century mathematicians, any abstract question, if it were to be considered relevant and worthy of consideration, had to be an abstraction from the physical world. It was characteristic of the age that general questions, such as which functions are allowable in mathematics, revolved around concrete physical cases, such as the motion of a taut string. It was therefore unremarkable that when Daniel Bernoulli weighed in on the debate and sided with Euler, he did so on the basis of purely physical considerations. He discussed a number of experiments showing that vibrating strings emit a whole range of tones and argued that this justified Euler's position about the strings' initial conditions.[26]

For Bernoulli, as for Louville some decades before, mathematical considerations could never be separated from physical reality. It was, in the last account, the world, the matter that composed it, and the laws that governed it, which determined whether a mathematical argument was correct or not. If it corresponded to physical reality, as Euler's position corresponded to Bernoulli's experiments, then the argument must be true; but if it did not, as was the case for d'Alembert's position, then the argument was false. For Bernoulli, as for Louville and other Enlightenment mathematicians, physical reality was the ultimate arbiter of mathematics.

From Geometry to Algebra

Like their predecessors, the leading mathematicians of the middle of the eighteenth century never doubted that mathematics was an abstraction of the physical world. D'Alembert said as much in his "Preliminary Discourse," and Euler expressed much the same sentiment when he wrote that "mathematics, in general, is the science of quantity, or the science which investigates the means of measuring quantity." He went on to define quantity as something "that is capable of increase or diminution" and gave as examples money, length, area, volume, time, mass, power, and speed.[27] This, however, does not mean that mathematics had remained unchanged from the time Johann Bernoulli used weights and strings to determine the shape of the catenary to the time his son invoked acoustic experiments in support of Euler's position on the vibrating string. Although Euler and d'Alembert viewed mathematics as an abstraction from physical reality, they were also intent on pushing this abstraction as far as they could. Whereas in the generation of the elder Bernoullis mathematical arguments had been dependent on specific geometrical constructions, mathematicians in the generation of Euler and d'Alembert believed that mathematics should be as abstract and general as possible. Algebra, being the most abstract mathematical field, was for them also the most general and most powerful.

It is easy to see why midcentury mathematicians, seeking to make their results as abstract and general as possible, preferred algebra to geometry. A geometrical construction, such as the one Johann Bernoulli had used to analyze the catenary, was necessarily dependent on the particular characteristics of the problem at hand. In order to arrive at the mathematical relationship describing the slope of the catenary, Bernoulli had to consider that the curve was defined as a hanging chain, that the forces at every point acted along the tangent of the curve, that the shape of the curve did not change if a section of it was removed, and so on. These are all features unique to the catenary and in no way apply to other curves such as the cycloid or the brachistochrone. An analysis of these curves would require altogether different constructions, dependent on their own unique features. This is inevitably the case with geometrical constructions, each of which is different from all others and therefore requires its own unique analysis. With each new problem a geometer has to start anew and establish the relationships that apply for this particular construction.

But where geometry is particular, algebra is abstract and general. A simple algebraic relationship such as $(a + b)^2 = a^2 + 2ab + b^2$ applies everywhere and always, regardless of the problem at hand, and the same is true of far more complex relationships, as long as one correctly adheres to the rules of algebra. To mathematicians of the mid-eighteenth century, such as d'Alembert and Euler, this was a clear sign of the superiority of an algebraic over a geometrical approach. If the problems of analysis could be expressed in algebraic terms, then their solution need no longer depend on the peculiarities of each separate question. Instead, one would be able to provide a single general algebraic solution that would apply to all individual cases. Such, they believed, was the power and the promise of algebra.[28]

D'Alembert's *Traité de dynamique,* first published in 1743 and considered by many his most original work, is emblematic of the direction mathematics was taking in the mid-eighteenth century. At the time of the *Traité*'s publication the French scientific community was still divided between Newtonians on the one side, and their Cartesian critics on the other. The former, deeply impressed with Newton's *Principia mathematica* of 1687, sought to disseminate the new Newtonian physics in France and expand its reach to areas that the great English master had not dealt with. The latter were suspicious of Newton's introduction of unexamined forces such as gravity into natural philosophy and championed the principles of "clear and distinct" reasoning introduced by Descartes a century before. Whereas the Newtonians saw the new physics as the only way forward, the Cartesians viewed it as a dangerous step backward to a "science" in which occult magical forces held sway.[29] By the 1740s, however, the Cartesians were fighting a rearguard battle, and the self-described Newtonians were clearly in the ascendant. Indeed, in his introduction to the *Traité* d'Alembert refers to the Cartesians, with some exaggeration, as "a sect which . . . scarcely exists today."[30]

In 1743 d'Alembert was 26 years old, a young member of the Paris Academy, and already an outspoken Newtonian. The *Traité,* his first major publication, was therefore an expansion of Newtonian mechanics, with a special emphasis on the rules governing the movement of objects through a resisting medium. At the same time it was also a critique of Newtonian mechanics and a proposal for an alternative, superior approach. D'Alembert's main issue with Newton's approach was its overreliance on unexamined "forces"—a criticism likely inspired by the much-derided Cartesians.[31] In place of a mechanics that relied on mysterious forces, d'Alembert proposed a mechanics that takes into account nothing but the observed fact of matter in motion. And unlike Newton's system, which focused on particular forces acting at specific points, d'Alembert hoped to formulate a mechanics based on general principles, applicable everywhere and always. Most significantly for our purposes, whereas Newton's *Principia* relied almost exclusively on a geometrical approach, d'Alembert's *Traité* was expressed almost entirely in the language of algebra.

There are many reasons why Newton had chosen to present his argument of the *Principia* in traditional geometrical form. Undoubtedly he considered that geometry was the most broadly accepted form of mathematical argument in his time, and not wishing to add another bone of contention to an already-radical argument, he chose geometry as the least controversial method. Newton may also have had a theological preference for the synthetic geometrical method, which began with first principles and built on them layer by layer. By following this logical process, one could follow in the footsteps of God as he laid down the laws of the universe. But quite apart from these strategic and philosophical reasons, there is no question that geometry lent itself very well to presenting Newtonian mechanics. The main argument of the *Principia* centered on the path moving bodies take when forces act on them at each and every point, ultimately showing that the orbits of the planets can be deduced from the universal law of gravity. Geometry, with its concrete material-looking diagrams, is naturally suited for presenting such an argument, with the curves representing trajectories, straight lines the actions of forces, unique points as centers of gravity, and so on. Geometry is, in a way, a natural language for a system of mechanics based on actual forces acting on material bodies in concrete locations.

It is hardly surprising that d'Alembert, who believed that generality is the highest virtue of a mathematical argument, was dissatisfied with Newton's approach. In place of the *Principia*'s concrete forces acting on bodies in specific locations, d'Alembert's *Traité* posited broad principles, valid everywhere and always. The culmination of the work was what came to be known as *d'Alembert's principle,* which dealt with the overall equilibrium of a mechanical system rather than with forces acting on bodies in a particular location and time. The actual movement of bodies could then be deduced from these principles without requiring the point-by-point analysis of the action of forces that had characterized Newton's approach. The principle was a source of great pride to its author, who years later still boasted of his accomplishment. "I believe," d'Alembert wrote in the *Encyclopédie* article "Mécanique," "that there is no problem in dynamics that cannot be resolved easily almost as a game by means of this principle, at least one can reduce it easily to an equation . . . It makes no use of actions or forces or any of those secondary principles."[32]

Doing away with forces, turning mechanics into a systematic game, and reducing its problems to equations was d'Alembert's goal in the *Traité de dynamique.* The mechanics of forces acting at a point was replaced in the *Traité* by the broad principles of the equilibrium of mechanical systems; and just as geometry had been the language of Newton's argument in the *Principia,* so was algebra the natural language of d'Alembert's games and equations. Abstract algebra, d'Alembert and his contemporaries believed, was the ideal tool for presenting the general principles of mechanics, transforming the study of concrete cases into a game, in which one equation follows necessarily from another.

Euler's Dilemma

A good illustration of the transition from a geometric to an algebraic conception of mathematical analysis can be found in Euler's demonstration of a simple and fundamental relationship of partial differential equations. The theorem states that if a function of two variables is differentiated first by one variable and then by the other, then the result is the same regardless of the order of differentiation. Known as *the equality of mixed partial differentials,* the result can be stated as follows in modern notation: if $z = z(x,a)$ is a function of two variables x and a, then $\frac{\partial^2 z}{\partial a \partial x} = \frac{\partial^2 z}{\partial x \partial a}$. Euler set out to establish this well-known relationship early in his career, in an unpublished treatise from the 1730s titled "On Differentiation of Functions Involving Two or More Variables," or simply "De differentiatione."[33] Significantly, the question here is much broader than the ones dealt with by Johann Bernoulli and his contemporaries a generation earlier. At issue is not the resolution of a particular physical problem, such as the shape of the catenary, but rather the general technique of the calculus itself. That Euler chose to address it is an indication of the shift from a geometrical approach that prevailed in the previous generation to more general and abstract mathematical methods. Nevertheless, Euler chose first to follow the example of his predecessors and provide a geometrical demonstration, likely inspired by the work of Nicholas Bernoulli (1695–1726), son of Johann and brother of Daniel.[34]

In Figure 2.6 from Euler's "De differentiatione," the curve AM represents the function $y = P(x,a)$ when the variable a is held constant, and the curve AN represents the same function P, this time when a is replaced by the constant value $a + da$, where da is an infinitesimal increment. Now, Euler argues, if $P(x,a)$ is differentiated while keeping a constant, then the differential at point P is the infinitesimal difference rm. The value of rm at every point, that is, the differential of $P(x,a)$ when a is constant, can be described by the function $d_x P(x,a) = S(x,a)$, which when applied to point P means that $S(P,a) = rm$. Similarly, if $P(x,a)$ is replaced by $P(x,a + da)$, then the differential at P is the infinitesimal increment ns, meaning that $S(P,a + da) = ns$. It follows from this that the differential of $S(x,a)$ when x is kept constant and a is varied is the difference $ns - rm$, or in more formal notation $d_a d_x P(x,a) = d_a S(x,a) = ns - rm$.

Next Euler differentiates $P(x,a)$ once more, this time varying a and keeping x constant. Because the curve ANn represents the function $P(x,a + da)$ and the curve AMm represents the function $P(x,a)$, the differential at point P is the difference between them, MN. The differential of $P(x,a)$ at every point when x is held constant and a is varied can be described as $d_a P(x,a) = T(x,a)$, which means that at point P, $T(P,a) = MN$.

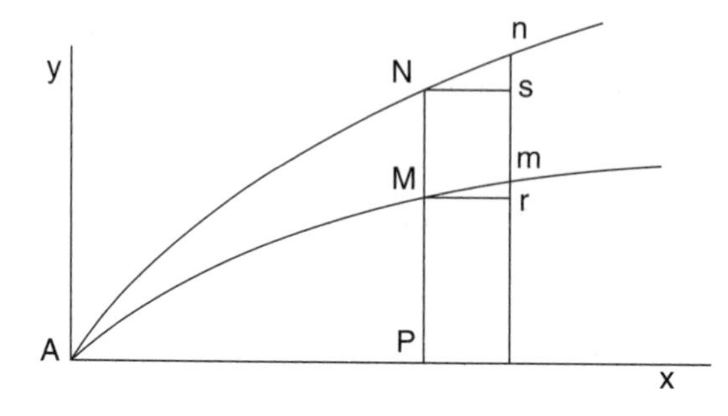

Figure 2.6. Leonhard Euler on the equality of mixed partial differentials in "De differentiatione." Steven B. Engelsman, *Families of Curves and the Origins of Partial Differentiation* (Amsterdam: Elsevier Science Publishers, 1984), 129.

Now, if an infinitesimal increment dx is added to P, we get $T(P + dx,a) = mn$. The differential of $T(x,a)$ at point P when a is held constant and x is varied is therefore $mn - MN$, or in more formal notation $d_x d_a P(x,a) = d_a T(x,a) = mn - MN$.

From the geometrical diagram it is clear, however, that $ns = nr - sr$, and $mr = nr - mn$, from which it follows that $ns - mr = mn - sr$. It is also obvious from the diagram that $sr = MN$, which means that $ns - mr = mn - MN$. As a result, $d_a d_x P(x,a) = d_x d_a P(x,a)$, which is a formal way of saying that that mixed partial differentials are equal, regardless of the order of differentiation. Q.E.D.[35]

Now although this proof would have appeared perfectly sufficient to an earlier generation of mathematicians, Euler is dissatisfied "since this demonstration is drawn from an alien fount," geometry. General algebraic results such as this, he implies, should be derived from algebra: now, he promises, "I will derive another one from the nature of such differentiation itself."[36] In this second demonstration Euler once more lets $P(x,a)$ be a function of two variables x and a.[37] If an increment dx is added to x, then P changes into Q, meaning that $P(x + dx,a) = Q(x,a)$. If an increment da is added to a, then P changes into R, meaning that $P(x,a + da) = R(x,a)$. And if both dx and da are added to x and a, respectively, then P changes into S, meaning that $P(x + dx,a + da) = S(x,a)$.

Now if we differentiate P while keeping a constant, we get

$$d_x P(x,a) = P(x + dx,a) - P(x,a) = Q(x,a) - P(x,a),$$

or, more simply,

$$d_x P = Q - P.$$

If we now differentiate $d_x P(x,a)$ while keeping x constant, we get

$$d_a d_x P = d_a(Q - P) = [Q(x,a + da) - P(x,a + da)] \\ - [Q(x,a) - P(x,a)] = S - R - Q + P.$$

Next Euler follows the same procedure but in reverse order. First differentiating P and holding x constant, he gets $d_a p = R - P$. Then, differentiating this result while holding a constant, he arrives at

$$d_x d^a P = d_x(R - P) = [R(x + dx,a) - P(x + dx,a)] \\ - [R(x,a) - P(x,a)] = S - Q - R + P.$$

And because $S - R - Q + P = S - Q - R + P$, Euler concludes that $d_x d_a P = d_a d_x P$. Q.E.D.[38]

The striking aspect of this demonstration to a modern reader is Euler's complete faith in the generality and reliability of the algebraic method. Euler is not at all concerned that his equations might be invalid for certain values or certain functions, as a present-day mathematician necessarily would be. Instead, he appears convinced that as long as he properly follows the rules of algebra in moving from one equation to the next, the truth of the result is guaranteed. This overarching faith in the truth value of algebraic transformations is typical of eighteenth-century mathematicians from Euler to Lagrange, who replaced the geometrical methods of their predecessors with a new algebraic approach. Algebraic expressions, they believed, were ultimately expressions of true relationships in the actual physical world. It followed that by moving from one expression to the next in accordance with proper algebraic rules, one was, in fact, revealing additional hidden harmonies that prevailed in the world. In this scheme there was no point in worrying too much about possible errors and the precise range of values in which an expression was valid. The reality of the world itself guaranteed the ultimate truth of algebraic expressions.

Euler's "De differentiatione" stands precisely on the cusp between two different mathematical ideals. On the one side is the geometrical model, which had been the standard of mathematical practice for millennia. On the other side is algebra, the abstract general method seeking to replace geometry as the foundation of mathematical practice. In "De differentiatione" Euler moves deftly from one to the other, making clear his preference for the new algebraic methods. For Johann Bernoulli in the 1690s, algebra was the handmaiden of geometry, its formulas facilitating the representation of essentially geometrical relations. But Euler in the 1730s wanted algebra to stand on its own and reduced geometrical arguments to "alien founts" that interfered with proper algebraic reasoning. Free from its geometrical moorings and representing true relations that existed in the physical world, algebra would become the new standard of mathematical practice.

Transferring mathematics from a geometric to an algebraic model was a primary goal of geometers in the age of Euler and d'Alembert. But although they managed to shift the focus of mathematical practice to algebra, they could never dissociate themselves completely from the geometrical forms of argument. D'Alembert's *Traité de dynamique,* for example, is typical of the hybrids of that period. It is a text dominated by general algebraic arguments, expressions, and equations, but it also contains its share of geometrical arguments and representations. In fact, to make his arguments d'Alembert uses no less than 79 different geometrical diagrams, which are not fundamentally different from those used by Newton and Johann Bernoulli 50 years before.

"No Figures Will Be Found in This Work"

The job of completing the transformation of mechanics into an algebraic science was thus left to d'Alembert's protégé, Joseph-Louis Lagrange, who in 1788 published his landmark study, *Mécanique analytique.* "I propose to condense the theory of this science [mechanics] and the method of solving the related problems to general formulas whose simple application produces all the necessary equations for the solution of each problem." This was precisely what d'Alembert had hoped for 45 years earlier, when he boasted in the introduction to the *Traité* that he had turned mechanics into an algebraic "game." But Lagrange proved truer to the algebraic creed than his mentor had been: "No figures will be found in this work. The methods I present require neither geometrical nor mechanical arguments, but solely algebraic operations subject to a regular and uniform procedure." In *Mécanique analytique* Lagrange effectively banished geometrical diagrams from mathematics, completing a century-long move from geometrical to algebraic forms of mathematical argumentation.[39]

It is instructive to compare a page from Newton's *Principia* with one from Lagrange's *Mécanique analytique,* almost exactly a century later. In proposition XI, problem VI, of book 1 of the *Principia* Newton is analyzing the forces acting on an object revolving in an ellipse. Naturally, the argument is accompanied by a geometrical diagram showing the elliptical path of the object, its tangent representing the direction of the object's movement at a given point, and the centripetal force acting on it and shaping its course. Similar diagrams that describe the motions of bodies under different conditions can be found on almost every page of the *Principia*. But in a comparable page from *Mécanique analytique* not a trace of a diagram is to be found. The entire argument proceeds through a succession of differential equations presented in pure algebraic form.

Whereas d'Alembert's *Traité* was a hybrid, combining algebraic aspirations with practical and intuitive geometrical techniques, Lagrange's *Mécanique analytique* was a purely algebraic construct. By transforming mechanics into an algebraic "game," in which any problem can be easily derived from a set of standard differential equations, Lagrange had fulfilled d'Alembert's promise from decades before.

In certain respects Lagrange's algebraic system is a testament to the profound transformation of mathematics during the eighteenth century. In 1692 Johann Bernoulli had struggled with only partial success to determine the geometrical shape of the catenary, the curve described by a hanging chain. His approach was materialistic, geometrical, and specific to the case at hand: he considered the weight of the chain, imagined what would happen if a section of it were removed, and quite possibly experimented to see if he was correct. His method for determining the tangent of the curve was wholly geometrical, and when he could not resolve the resulting differential equation, he turned back to geometry, relating the catenary to the area under a familiar conic section, the hyperbola. But less than a century later Lagrange had completely erased not only material auxiliaries such as weights and strings, but any reference to geometry as well, replacing it entirely with abstract algebraic relations. *Mécanique analytique* was a work of mathematics immeasurably broader, more general, and more abstract, as well as more powerful, than Johann Bernoulli's geometrical manipulations. In the course of one century mathematics—a field famous for its long-term stability—had evolved almost beyond recognition.

But if a comparison of mathematical practices in the late seventeenth and eighteenth centuries points to some stark contrasts, it also reveals that profound underlying assumptions had changed not at all in a century of work. Mathematics for Lagrange was fundamentally about the underlying structure of the physical world, just as it had been for the Bernoullis, d'Alembert, and Euler—so much so that throughout the century mathematical discussions and developments were centered on issues that would today fall under the purview of physics. In the early years, around 1700, the questions were specific and dealt with chains and the paths of rolling balls. In the middle of the century mathematicians such as Maupertuis, d'Alembert, and Euler dealt with broader questions, such as the equilibrium of mechanical systems and the principle of least action. By the end of the century Lagrange's work demonstrated how the complete science of mechanics could be expressed by sets of abstract differential equations. But through all these changes the ultimate subject matter of mathematics remained unchanged: it was the deep structure and order of the physical world.

The Crisis of Natural Mathematics

The belief of Enlightenment geometers in the physical roots of mathematics, on the one hand, and their commitment to ever-increasing abstraction and generalization, on the other, created an inner tension in the field that increased over time. By the end of the eighteenth century a growing anxiety about the field and its future was palpable among the leading mathematicians. As they were well aware, at the far end of abstraction loomed the danger of complete dissociation from the physical world, of a discipline that was so successful at removing itself from the contingencies of physical reality that it could not find its way back. Such a field would be a labyrinth of self-referentiality that would point to nothing outside itself. It would confirm the charges of critics such as Diderot and Buffon that mathematicians had climbed so high up their mountain of abstraction that they could no longer find their way back to reality. To mathematicians, this kind of mathematics was a nihilistic nightmare that must be avoided at all costs. Ever-greater abstraction was indeed the guiding principle of the field throughout the long eighteenth century and gave mathematics coherence and direction; but too much abstraction was extremely dangerous and posed the risk of undermining the very foundations of the field.

Torn between the imperative to greater abstraction and its inherent dangers, Enlightenment mathematicians were caught in a high-stakes dilemma. The shift toward algebra had proved extremely fruitful and had brought the field its greatest triumphs, but any further move in that direction could destroy the field altogether. Lagrange was the most adventurous of the grands géomètres, willing to push the field toward extremes of algebraic abstraction, whereas his more cautious contemporaries hesitated to join in. But proud as he was of his achievements, Lagrange was not immune to the anxieties of his colleagues about the future of mathematics. "The mine is almost too deep already, and unless new seams are discovered, it will be necessary to abandon it sooner or later," he wrote to d'Alembert in 1781. It might soon be time, he suggested, to concentrate on physics and chemistry—fields that unlike mathematics never strayed far from their roots in the material world.[40]

From Johann Bernoulli to Joseph-Louis Lagrange, mathematics was worldly and the world was mathematical. On the one hand, mathematics was a generalization and abstraction of the material relations that prevailed within the physical world; on the other, the world was a concrete expression of general mathematical relations.[41] Lagrange elegantly expressed this relationship when he wrote that his accomplishment in *Mécanique analytique* was to have made mechanics into a branch of mathematical analysis. D'Alembert expressed it more systematically in the "Preliminary Discourse," writing that each branch of mathematics represented a different degree of abstraction from physical reality, with algebra being the "farthest outpost" one could reach "without leaving the material universe altogether." In turn, he argued, the mathematical relationships one discovers at this final frontier of abstraction reveals to us the hidden relationships that prevail within the material world.[42]

If mathematics was indeed a "natural" science, concerned with the hidden harmonies of the physical world, then those best suited to practice it were, of course, "natural men," men like the stodgy Euler, simple and unassuming, his feet planted firmly on the ground and forever suspicious of metaphysical flights of fancy; or even better, men like d'Alembert, the penultimate Enlightenment mathematician and, to his admirers, the embodiment of Rousseauian natural simplicity. D'Alembert had fashioned himself as a man without parentage, a child of nature who never strayed far from her simple ways. Unlike other men, d'Alembert, to his admirers, never lost his natural simplicity and looked at the world with a child's innocent curiosity. A man such as d'Alembert, they implied, has an unmediated connection to nature, unencumbered by the artifices of social conventions and interests. And who is better positioned to reveal nature's hidden harmonies than a man who never really left nature behind? Simply put, in the age of Enlightenment the ideal mathematician was a true natural man; his field, mathematics, was a true science of nature.

So things stood as the eighteenth century drew to a close and the nineteenth century dawned amid the turmoil of war and revolution. A new generation of "natural" mathematicians who were trained in the final decades of the old century and reached maturity under the Revolution and the Empire continued and improved on the traditions handed down to them by d'Alembert, Euler, and Lagrange. Gaspard Monge (1746–1818), Pierre-Simon Laplace (1749–1827), Lazare Carnot (1753–1823), and Joseph Fourier (1768–1830), to name only the most illustrious of that generation, were true heirs of the grands géomètres of the Enlightenment in both their lives and their professional practice. Even more than their predecessors, they were men of action, high officials of state under the revolutionary regime and Napoleon, and personal friends of the greatest men of the realm. Their mathematics too marked them as the children of d'Alembert, focused as it was on physical problems and seeking to unveil the hidden structures of the world around us.

But on the streets of Paris in those same years a young mathematician lived and breathed who resembled these great men not at all. An angry young man snubbed by the scientific establishment, he lived on the margins of society, devoting his energies to a doomed revolutionary cause. He was a man without power and without connections, emotionally troubled and ill at ease in the world, who died violently at the tender age of 20. His mathematical work, furthermore, was so far removed from the physical questions that dominated the field that academicians threw up their hands in despair at trying to understand what the young man was up to. And yet, not only did the mathematical achievement of this young misfit come to outshine that of nearly all his established contemporaries, but the fame of his short and tragic life story far eclipsed the brilliant careers of Laplace, Fourier, and their colleagues. Astoundingly, his life story of failure, misery, and early death became a model of the truest and purest mathematical life for generations to come. His name was Évariste Galois.

II

HEROES AND MARTYRS

3

A HABIT OF INSULT: THE SHORT AND IMPERTINENT LIFE OF ÉVARISTE GALOIS

Truth and Lies

Évariste Galois was a young mathematical genius in early nineteenth-century Paris. Despite his obvious brilliance and groundbreaking solutions to questions that had resisted the efforts of the greatest mathematicians, Galois did not receive the recognition he deserved from the mathematical establishment of his day. Twice he was refused admission to the École polytechnique, and when he began his studies in the less prestigious École normale, he was expelled after less than a year. Convinced of the importance of his work, Galois submitted his discoveries for evaluation by the Paris Academy of Sciences, but the memorandum was promptly lost by Augustin-Louis Cauchy, the leading mathematician of the day. Twice more Galois attempted to interest the academy in his work, only to be turned away without a hearing.

Disgusted and disillusioned, Galois turned to radical politics and was soon arrested for sedition, landing in prison for several months. Shortly after his release he fell in love with a mysterious young woman named Stéphanie, but the affair went badly and he was challenged to a duel by a rival. Knowing that he might not survive the dawn, Galois spent the night before the duel fiercely jotting down his latest and most profound mathematical insights. "I have no time!" he scribbled in the margins. Tragically, Galois' premonition proved true. On the morning of May 30, 1832, he was shot in the stomach by his opponent and left to die on an empty Paris street. Galois died five months short of his twenty-first birthday, but his mathematical testament of his last night survived and bequeathed to mathematics an entire new field of study: group theory, a central pillar of modern algebra.

Such is the story of the short life and early death of Évariste Galois as told by its most successful popularizer, the mathematician and author Eric Temple Bell.[1] It is likely the most famous of all mathematical biographies and has been cited as a source of inspiration by generations of mathematicians and physicists, including John Nash (of *A Beautiful Mind* fame) and Freeman Dyson.[2] The mathematician James R. Newman, the physicist Leopold Infeld, and the astronomer Fred Hoyle, all leaders in their fields, each published their own varying accounts of the life of Galois, and in 1948 Alexandre Arnoux directed the movie *Algorithme* based on Galois' life.[3] More recently a biography by Laura Toti Rigatelli (1993), an account by astronomer Mario Livio (2005), and a full-length novel by Tom Petsinis titled *The French Mathematician* (1997) attest to the continuing appeal of Galois' story.[4] The life of Galois "is the first and primary story that novice mathematicians learn about being a mathematician," a recent web article proclaims, and my own experience fully supports this observation.[5] As an undergraduate majoring in mathematics at the Hebrew University of Jerusalem in the 1980s, I heard the story repeatedly from my professors, who presented it as something of an introduction to the serious study of mathematics.

The reasons for the continuing fascination with Galois are not hard to fathom. Galois seems like a figure out of a novel: young, handsome (to judge by the two surviving sketches of him), and boundlessly idealistic, quite apart from being a mathematical genius. Our hearts go out to young Galois, who was always uncompromisingly true to himself without ever stopping to consider the practical consequences of his actions. Whether in mathematics, politics, or love, his devotion to truth and honor knew no limits. Galois suffered and died for following his heart with such desperate determination, and his example certainly does not recommend itself as a desirable approach to life. But who among us, who exist in a world of compromise and shades of gray, does not wish that for a single moment they could live life as brightly and fully as Galois did in his twenty years?

But there are other reasons for the popular fascination with Galois as well. Like other stories that exhibit such constant appeal through generation after generation, the tale of the life and death of Galois is first and foremost a morality tale. Put simply, it is a story of innocence crushed by the ugly exigencies of worldly life. Galois was that truly rare individual, pure of heart and pure of mind, who refused to make the small compromises that make up the daily lives of well-adjusted people. Comfortable in their institutional positions and compromised by their worldly lives, Cauchy and his colleagues refused to acknowledge the brilliant mathematical gem that was presented to them. They turned away the gift of the true genius, driving him to despair and early death.

Although Galois' case is undoubtedly extreme, many, if not most, of us will recognize the sentiment expressed in his tale. The story of Galois appears to us as a perpetual warning against the arrogance and complacency that can sometimes be found in staid and established institutions such as the Academy of Sciences. Satisfied that no outsider could possibly contribute to the progress of science, the academicians gave short shrift to a brilliant contribution that was right before their eyes. But true genius, Galois' story suggests, lies not in the stuffy halls of the academy but in the pure and innocent mind of a young man, whose brilliance and love of truth transcended his dire physical circumstances. The conflict between true genius and institutional hard-heartedness inevitably leads to disaster.

The tale of the life and death of young Galois combines the personal appeal of a vivid personality with a moral lesson that applies to us all. Nevertheless, it suffers from a single fatal flaw: it is not true. As numerous modern studies have shown, essential parts of the story are at least highly exaggerated, if not wholly fabricated. In truth, the members of the mathematical establishment were far from hostile to Galois and were much more receptive to his work than the oft-told legend suggests. And although there is no question that Galois' suffering was great, it is also clear that the young mathematician's own intransigence and paranoia were responsible for a good portion of his troubles.

The most glaring fallacy in the legend of Galois is also its most dramatic element, and the one that has made the deepest impression on succeeding generations. That is the tale that the young genius wrote down his complete mathematical testament on the very eve of his death, and that these hastily jotted notes became the basis for the field of group theory. The fact is that by the time of his death Galois had published no less than five separate articles in the prestigious journals *Annales de Mathématiques* and the *Bulletin des Sciences Mathématiques.* On his last night Galois did indeed compose a long letter to his friend Auguste Chevalier that summarized the main points of his discoveries. But it was the published articles, along with the memoir he submitted for the academy's grand prize, that are truly responsible for Galois' mathematical reputation.

Further inspection reveals more cracks in the facade of the Galois legend. According to the popular version, Galois had submitted a memoir to the academy that was either ignored or lost by Cauchy, the most prominent mathematician of the age. In fact, a letter by Cauchy to the academy requesting time to present Galois' work shows that he considered it very seriously, although he did not in the end present it at the appointed time. A letter that appeared in the journal *Le Globe* at the time of Galois' arrest states that in fact "M. Cauchy had conferred the highest praise on the author," implying that far from ignoring Galois, he may well have encouraged him in his work.[6]

A similar picture emerges from Galois' other interactions with the academy. In 1830 Galois presented a memoir to the academy for consideration for the grand prize in mathematics. Unfortunately for the young mathematician, the academician assigned to review his work, Joseph Fourier, died before submitting his report. When Galois' manuscript was not found among Fourier's papers, he was summarily dropped from consideration for the prize. Despite this setback, the following year Siméon-Denis Poisson, who was himself a leading academician, approached Galois and asked him to submit a detailed memoir on his work to the academy. Galois did so, producing his most detailed presentation of what would become *Galois theory*. Poisson's report, when it came some months later, was undoubtedly disappointing to Galois: "We have made all possible efforts to understand M. Galois' proof," Poisson wrote, adding that "his reasonings are neither sufficiently clear, nor sufficiently developed for us to give an opinion in this report." Nevertheless, Poisson did not reject Galois' claims but rather was hoping for a clearer and more complete presentation. "One should therefore wait for the author to publish his work in its entirety to form a definitive opinion," he concluded.[7]

Overall, Galois' relationship with the mathematical establishment was certainly difficult. He had suffered setbacks and bad luck, including Fourier's untimely death, Cauchy's failure to deliver on his promised presentation to the academy, and Poisson's lukewarm report on his memoir. Nevertheless, at the tender age of 20 Galois was in regular contact with the leading mathematicians in the world, who took his work very seriously. Cauchy reputedly encouraged him, Poisson solicited a memoir from him, and both were apparently hoping for more complete and more accessible presentations. He was also publishing regularly in journals where his work appeared side by side with articles by Cauchy and other top mathematicians. How many 20-year-old students could claim to have accomplished as much or to have gained as much recognition in so short a time? Although he viewed himself as a victim of systematic persecution by a narrow-minded establishment, in truth Galois was considered a young mathematician of great promise by the leading mathematicians of the time.

"A Habit of Insult"

Galois, however, had a talent for making things difficult for himself. Already as a student at the Lycée Louis-le-Grand, where he spent most of his teenage years, Galois was recognized not only for his high intelligence but also for his mercurial personality. According to his school reports, a promising start in his early years at the school soon turned sour. By the time he was 16, one puzzled professor noted a "character which I do not flatter myself I understand every trait," while another already expressed outright hostility: "There is no trace in his tasks of anything but of queerness and negligence." "The furor of mathematics possesses him," noted a third: "He is losing his time here and does nothing but torment his masters and gets himself harassed with punishments."[8] With hindsight, knowing Galois' extraordinary mathematical talent, we may be inclined to excuse his impatience with the school curriculum. Nevertheless, this is hardly an attitude that is likely to make friends and smooth one's way in the world.

In 1829, after being rejected for the second time by the École polytechnique, Galois enrolled in the less prestigious École normale. There he became an ardent republican and wasted little time before entering into a bitter conflict with none other than the director of the school, Joseph-Daniel Guigniault. During the turbulent days of July 1830, when pitched battles were fought on the streets of Paris, the students of the École polytechnique were at the forefront of the revolution, leading republican forces against loyalist troops. Guigniault, however, locked the students of the École normale indoors to prevent their taking part in the events, and Galois, despite earnest attempts to scale the walls, missed his chance at revolution. He never forgave Guigniault, and their deteriorating relationship soon reached a crisis. In December 1830 Guigniault published a letter criticizing a liberal teacher at the Lycée Louis-le-Grand, and Galois took up the teacher's defense. In a letter in the student newspaper *La Gazette des Écoles,* Galois accused Guigniault of having a "narrow outlook and ingrained conservatism" and promised to "unmask this man."[9] Although the newspaper's editors removed the signature from this inflammatory letter, it did not take long for Guigniault to track down its author. Within a month, in January 1831, Galois was expelled from the École normale.

Galois' behavior in this affair followed a pattern that became familiar throughout his short life. With the hindsight of nearly two centuries we may feel more charitable toward poor Guigniault, who was caught up in dramatic events far beyond his control. He hedged his bets as best he could between royalists and revolutionaries, resisted the transformation of his school into an armed republican camp, and did his best to stay on the sidelines and wait for the storm to pass. To Galois, however, this rather reasonable policy was nothing short of treason. The choice for Galois, then as always, was simply between right and wrong. Any attempt at compromise was a betrayal of truth in the interest of falsehood and could never be condoned. The justice of the republican cause was to Galois as certain and incontrovertible as the truths contained in his mathematical work. Both had to be defended, and Galois was ready to take on the fight, whatever the consequences.

That the same pattern of provocation characterized Galois' relations with established mathematicians is clear from the testimony of Sophie Germain (1776–1831), the only woman in the exclusive society of higher mathematics of the time. In a letter to the mathematician and academy member Guillaume Libri (1803–1869) on April 18, 1831, she lamented what she regarded as the sad state of mathematics in Paris:

> Decidedly there is misfortune concerning all that touches upon mathematics. Your preoccupation, that of Cauchy, the death of M. Fourier, have been the final blow for this student, Galois who, in spite of his impertinence, showed signs of a clever disposition. All this has done so much that he's been expelled from l'École Normale. He is without money, and his mother has very little also. Having returned home, he continued his habit of insult, a sample of which he gave you after your best lecture at the Academy. The poor woman fled her house, leaving just enough for her son to live on, and has been forced to place herself as a companion in order to make ends meet. They say he will go completely mad. I fear this is true.[10]

Germain's casual reference to Galois indicates that he was well known to Libri and his colleagues, as was his outrageous behavior. Here we get a glimpse of Galois attending a lecture by Libri at the academy and taking center stage by heaping scorn on the speaker. We also get a glimpse of his sad family life, and although we have no corroborating evidence that Galois' mother "fled her house," it is not on the whole surprising if Galois' abrasive behavior did not limit itself to established mathematicians and school directors.

Out on the streets of Paris after his expulsion from the École normale, Galois became active in the Artillery of the National Guard, a radical republican militia that was officially disbanded by the new government of King Louis-Philippe. When nineteen "patriots" were acquitted of serving in the outlawed militia in April 1831, radical republicans celebrated their release in a Paris tavern. "It would be difficult to find in all of Paris two hundred guests more hostile to the government than those," wrote the novelist Alexandre Dumas, who was in attendance. In his memoirs Dumas described the events of the night:

> Suddenly . . . the name Louis-Philippe, followed by five or six whistles caught my ear. I turned around. One of the most animated scenes was taking place fifteen or twenty seats from me. A young man, holding in the same hand a raised glass and an open dagger, was trying to make himself heard. He was Évariste Galois . . . one of the most ardent republicans. All I could perceive was that there was a threat, and that the name of Louis-Philippe had been pronounced; the intention was made clear by the open knife.

Dumas was possessed of a keen instinct for self-preservation that was sadly lacking in Galois himself. Noting that "this went far beyond the limits of my republican opinions," he made a quick exit by jumping out a window into the garden and heading home. "It was clear," he wrote, "this episode would have its consequences." "Indeed, two or three days later Évariste Galois was arrested."[11]

During his trial, which took place two months later, Galois explained that he had indeed brandished the knife with his raised glass, but that his toast was "to Louis-Philippe, *if he betrays us.*" Unfortunately, he said, the last words were drowned by the whistles and clamor of those who thought that he was toasting the health of the king. Because Galois, in response to questioning, also insisted that he saw no reason to believe that Louis-Philippe would not betray him and his comrades, this could hardly be considered a sterling defense. The sympathetic judge, who had evidently taken the measure of the young firebrand, cut short Galois' closing statements before he could do any more damage to his case. He also acquitted him of all charges.[12]

Free again after two months of incarceration, Galois managed to stay out of trouble for less than a month. On Bastille Day, July 14, 1831, he led a crowd of 600 ardent republicans in a march across the pont Neuf. Their destination was the place de Grève, in front of city hall (the Hôtel de Ville), where they planned to plant a liberty tree. As every Parisian knew, the place de Grève was the site of the pillory and of executions before 1789, and in all of Paris it was the place most associated with the ancien régime and its alleged crimes. By planting a liberty tree in this location, the republicans were putting the government on notice that the revolution was not over, and the crimes of monarchy were not forgotten. Louis-Philippe, they implied, might well suffer the same fate as his Bourbon predecessors.

This was one of many such events planned by republicans that day all over Paris in commemoration of the liberty trees that had been planted on Bastille Day 40 years before, at the height of the Revolution. The police took swift preventive action: on the night of July 13 they swept through the city and arrested those they considered the ringleaders of the events. Galois was spared, probably because he took the precaution of staying away from his room that night. The following day, however, as he and his friend Ernest Duchâtelet were leading the marchers across the pont Neuf, the two were taken into police custody. At the time of his arrest Galois was dressed in the uniform of the outlawed Artillery of the National Guard and was carrying a loaded carbine, two pistols, and a knife.[13]

It was more than two months before Galois was brought to trial, and during this entire time he was jailed without charge. The chances of a good outcome for him and his friend were not improved when Duchâtelet drew a caricature of King Louis-Philippe's head shaped as a pear on their cell's wall. He added a guillotine next to it, with the inscription "Philippe will carry his head to your altar, O Liberty!"[14] When Galois was finally brought to trial on October 23, he received what was considered a harsh sentence: six months in Sainte-Pélagie prison.

If the experience of incarceration in a dreary Parisian prison would have served as a warning against provocative behavior to most young prisoners, this was not the case for Galois. On July 29, just two weeks after his arrival at Sainte-Pélagie, young Galois was a member of a delegation sent by the prisoners to negotiate with the chief warden. The occasion was a shot fired from outside the prison's walls that had injured one of the prisoners, and Galois did not hesitate to accuse one of the guards of being the shooter and also to insult the chief warden. It was almost a replay of Galois' confrontation with his old nemesis, Director Guigniault of the École normale, and the results were just as grave for the young mathematician: he was thrown into the dungeon. This time, however, Galois received more support from his peers: "Galois in the dungeon!" they protested. "Oh, the bastards! They have a grudge against our little scholar." The prisoners then took over Sainte-Pélagie, and order was not restored until Galois was released from the dungeon the following day.[15]

Most of what we know of Galois' time in prison comes from the memoirs of François-Vincent Raspail, scientist, revolutionary, and ultimately statesman, who is honored today with a wide Paris boulevard and a metro stop on the Left Bank. Raspail refused to receive the Cross of the Legion of Honor for his scientific work from Louis-Philippe, and in 1831 he was serving time as a political prisoner in Sainte-Pélagie. At his own trial Raspail had declared that the king "should be buried alive under the ruins of the Tuileries," showing that Galois was not the only republican fond of issuing grandiose, provocative proclamations. But Raspail's long career shows that he was able to mix his heartfelt beliefs with a sobering sense of reality and its possibilities. Galois, in contrast, believed every word he uttered without reservation and was ready to back it up without regard for the consequences.

Raspail understood this about Galois, and in his *Réforme pénitentiaire: Lettres sur les prisons de Paris,* published eight years after the events, he recounted what happened when Galois, who had little experience with alcohol, was challenged by fellow prisoners to a drinking contest. For Galois, Raspail explained, refusing the calling "would be an act of cowardice, and our poor Bacchus has so much courage in his frail body that he would give his life for the hundredth part of the smallest good deed." Galois then drank glass after glass "like Socrates courageously drinking the hemlock" until he collapsed.[16]

Galois, as Raspail saw clearly, was incapable of stepping back from a challenge to his honor, even in a matter as trivial as alcohol consumption. In moments of clarity Galois also seemed to realize this about himself. After a similar occasion, in which he emptied a bottle of brandy, he confided his inner torment to the older Raspail: "What is happening to my body? I have two men inside me, and unfortunately I can guess who is going to overcome the other . . . See here! I do not like liquor. At a word I drink it, holding my nose and get drunk." Despite this flash of self-knowledge, Galois felt helpless to resist the inner fires that were driving him to disaster. With eerie precision he went on to predict his own death: "I do not like women . . . and I tell you, I will die in a duel on the occasion of some 'coquette de bas étage.' Why? Because she will invite me to avenge her honor, which another has compromised." Following this prophecy, Galois announced that he must kill himself, and he would have done so, according to Raspail, if the prisoners had not flung themselves on him and pulled the weapon out of his hand.[17]

Duel at Dawn

The circumstances of Galois' final duel and death have been the focus of an enormous amount of speculation, beginning immediately after the event. The mathematician's younger brother, Alfred Galois, believed all his life that Galois was the victim of a police setup, aimed at eliminating a vocal republican leader. In later years the claim was taken up by Leopold Infeld, who in his biography of Galois argued that the mysterious "Stéphanie" was a police agent who deliberately provoked the duel.[18] Fred Hoyle, in a partial inversion of Infeld's theory, argued that Galois' ability to carry out complex calculations in his head made his republican comrades suspect that he was a police agent. They deliberately provoked the duel to eliminate him.[19] More recently, Laura Toti Rigatelli has argued that the duel was orchestrated by Galois himself, who decided to sacrifice his life in order to arouse the masses to revolution.[20] Others have speculated extensively over the identity of Galois' opponent in the duel: Tony Rothman, following a comment by Dumas, claims that Galois was killed by Pescheux d'Herbinville, a fellow radical and member of the Artillery of the National Guard. Mario Livio argues that Galois' opponent was none other than Ernest Duchâtelet, the very man who marched side by side with Galois across the pont Neuf on Bastille Day. I do not claim to choose between these rival versions, although certainly the more extreme conspiracy theories appear far-fetched and have not been widely accepted. There is no solid reason to believe that the duel was anything but what it appeared to be—a confrontation over a young woman's honor.[21]

Here is what is known about the events that transpired in the final months of Galois' life: In the spring of 1832 a cholera epidemic broke out in Paris, and Galois was transferred from Sainte-Pélagie and placed on parole at a convalescent home known as Sieur Faultrier. There he met and fell in love with 16-year-old Stéphanie Poterin du Motel, whose family was residing in the same house.[22] Fragments of two letters from Stéphanie that Évariste copied indicate that the affair, such as it was, did not go well, and that Stéphanie was trying to distance herself from the intense young mathematician. Galois was devastated. Dramatic as ever, he wrote desperately to his friend, Auguste Chevalier, on May 25: "How can I remove the trace of such violent emotions as those which I had experienced? How can I console myself when I have exhausted in one month the greatest source of happiness a man can have? When I have exhausted it without happiness, without hope, when I am certain I have drained it for life?"[23]

It appears that at some point in their interactions the ardent Galois had transgressed the bounds of propriety and given offense to Mlle. du Motel. To her defense she called two men, both reportedly "patriots" and friends of Galois, who challenged the young mathematician to a duel. In a letter "to all republicans" he wrote on his last night, May 29, Galois referred to Stéphanie as "an infamous coquette"—reminiscent of the "coquette de bas étage" of his prophecy to Raspail, but very different in tone from his letter of only four days before, when he bemoaned the loss of his life's true love. He went on to claim that he was taking part in the duel as a last resort, after having exhausted every means of avoiding the confrontation. "I repent having told a baneful truth to men who were so little able to listen to it calmly," he added. "Yet I have told the truth. I take with me to the grave a conscience clear of lies."[24] Perhaps we should not be overly surprised at the failure of peace overtures that involved Galois confronting his challengers with a "baneful truth" on the grounds that his conscience would not allow him to lie; and it is, on the whole, not unexpected that when confronted with this "truth," the two men found themselves "little able to listen to it calmly." But nothing could be more characteristic of Galois, who saw a world composed only of truth and falseness, and for whom compromise and textured shades of gray were utterly alien.

Galois wrote two more letters on his last night. In one, addressed to "two republican friends," N.L. and V.D., he asserts once more that he had tried in vain to reach a compromise, but again insists on the truthfulness of his claims. In fact, he charges his correspondents to testify "whether I am capable of lying, even on such a trivial subject."[25] Clearly he was not, even at the price of risking his life over what he went on to call "a miserable piece of gossip." The other letter is addressed to Auguste Chevalier and contains a summary of his mathematical work. It is in large part a reprise of his last memoir to the academy, although it contains a few additional theorems. It is this letter that gave rise to the legend that Galois wrote down the theory of groups in its entirety on the night before his death.

Galois met his adversaries in the early morning of May 30 in Gentilly in southern Paris. According to a report in the Lyon newspaper *Le Précurseur* several days later, the duelists left the outcome of the encounter to pure chance by having only one of the pistols loaded.[26] When the shot rang out, Galois was struck in the belly and fell to the ground. Several hours later he was brought to the nearby Cochin hospital, where he was joined by his 18-year-old brother Alfred. When both brothers realized that the end was near, Évariste spoke his last words to his brother: "Don't cry," he said; "I need all my courage to die at twenty." At ten in the morning of May 31, 1832, Évariste Galois passed away.

What are we to conclude from the dramatic life and death of the young French mathematician? No doubt his life was not an easy one. From his repeated rejection by the École polytechnique to the questionable treatment of his memoirs by the members of the Academy of Sciences, Galois had to deal with substantial setbacks and unnecessary obstacles that were strewn along his way. Nevertheless, the legend of Galois, which presents him as the victim of persecution by a closed-minded mathematical establishment, clearly misses the mark. One could probably forgive the gentlemen at the academy if they dismissed the young firebrand, who by the age of 20 had managed to get expelled from school, had spent the better part of a year in prison, and had acquired the reputation of a dangerous radical. But they did not. At the time of his death Galois was viewed as a young man of great promise in the highest mathematical circles.

What is clear from all this is that Galois' worst enemy was neither Cauchy nor Guigniault, nor even the rival who eventually killed him. It was, unquestionably, Galois himself. Already in his days at the Lycée Louis-le-Grand we hear of his dismissive attitude toward the school curriculum, and it seems very likely that his arrogant behavior during his oral examinations had much to do with his rejection by the École polytechnique (a legend that has him hurling an eraser at a pedantic examiner is unsubstantiated). At the École normale he immediately locked horns with the school's director, essentially forcing his own expulsion, and he acted in almost exactly the same manner with the chief warden at Sainte-Pélagie the following year. His political provocations, from toasting the king with an open dagger to leading a Bastille Day march while armed to the teeth and dressed in the uniform of a banned militia, left authorities little option but to send him to prison. He did no better in his relations with the academy, where instead of cultivating his promising connections he preferred to insult its members in public. And when he was challenged to a duel by men who were fellow radicals and reportedly his own friends, he was far from apologizing for his actions or seeking a compromise, insisting instead on confronting them with the "baneful truth."

There is a grim consistency in all this. Convinced from the start that the decks were stacked against him, Galois turned this conviction into a self-fulfilling prophecy by his own actions. He forever saw himself as the victim of malevolent machinations, whether by school authorities, the government of Louis-Philippe, or the mathematicians of the academy. He responded with self-righteous rage, which he undoubtedly considered a justified defense or simply "the truth," but which to others appeared as senseless provocation. Ultimately he succeeded in alienating many of those who could have been his friends and supporters. Even more tragically, and just as he had predicted, Galois managed to corner himself into a confrontation from which he could not back out. Laura Toti Rigatelli may be wrong in her theory that the duel was a hoax contrived by Galois as the means of his own suicide, but the more general notion, that the duel was the inevitable final station in a reckless path of self-destruction, seems perfectly valid. If any person was ultimately to blame for the short and tragic life of the brilliant young mathematician, it was, inescapably, Évariste Galois himself.

"Martyred by His Genius": The Posthumous Life of Évariste Galois

The historical Galois, the young radical with a giant chip on his shoulder, the brilliant but obnoxious mathematician whose behavior alienated friend and foe alike, was soon forgotten. Nowhere in the subsequent telling of his life do we learn about his "habit of insult" or of his driving his widowed mother out of her home. In its place there emerged quite a different character, the Galois of legend. This Galois was a gentle being, a pure soul of rare brilliance, oppressed by the stupid and corrupt authorities who controlled his fate. Far from being a provocateur, the Galois of legend was a victim, a casualty of a crass and uncomprehending world.

The recasting of the young mathematician as a victimized martyr began even during Galois' lifetime, spurred by Galois himself and his republican friends. Already on the day of Galois' trial for his knife-aided toast to the king, an article in the Saint-Simonian republican daily *Le Globe* described Galois' genius and recounted at length his fruitless efforts to interest the academy in his work. Galois' submission to the grand prize competition, the author explained, overcame fundamental problems in the solvability of algebraic equations that even the great Lagrange could not overcome. Cauchy himself, we learn, "had conferred the highest praise on the author on this subject."

> But what did it matter? The memoir was lost, the prize was awarded without the young savant being able to participate in the competition. In reply to a letter to the Academy from young Galois, complaining about the negligent treatment of his work, all that M. Cuvier could write was: "The matter is very simple. The memoir was lost on the death of M. Fourier, who had been entrusted with the task of examining it."[27]

This setback, the article explained, was then followed by another, when at Poisson's request Galois submitted another memoir to the academy, "the result being that for more than five months its wretched author has been waiting for a kind word from the Academy."[28]

The author of the article was probably Galois' friend Auguste Chevalier or his brother Michel Chevalier, both of whom were committed Saint-Simonians and contributors to *Le Globe*.[29] In the article's account, written nearly a year before Galois' death, we already see the basic outlines of the "Galois legend" as it was to develop over the years. Galois here is a gentle genius who meekly submits his mathematical gems to the judgment of his superiors in the full expectation that they will acknowledge their worth. Instead, he is confronted with a wall of institutional indifference that dismisses his work and crushes his hopes.

Paradoxically, it was Galois himself who proactively and efficiently did more than anyone else to propagate his image as a passive victim of circumstances. In his private prison confessions to Raspail he could and did acknowledge the self-destructive demons that were driving him to disaster. But in letters to his friends and in a preface to his collected works that he hoped to publish after his release from prison, the Galois of legend is clearly taking shape. In this work, also written during his stay in Sainte-Pélagie, Galois rails against the "greats" of the world, who have teamed up to crush the hopes of a brilliant young mathematician. His resentment and sense of persecution by a nameless establishment are palpable:

> Firstly, the second page of this work is not encumbered by surnames, Christian names, titles, honors, or the eulogies of some avaricious prince whose purse would have opened at the smoke of incense, threatening to close when the incense holder was empty. Neither will you see in characters three times larger than those in the text, homage respectfully paid to some high-ranking official in science, a thing thought to be indispensable (I should say inevitable) for someone wishing to write at twenty. I tell no one that I owe anything of value in my work to his advice or encouragement. I do not say so because it would be a lie. If I addressed anything to the important men of science or of the world (and I grant the distinction between the two [is] at times imperceptible), I swear it would not be thanks. I owe to important men the fact that the first of these papers is appearing so late. I owe to other important men that the whole thing was written in prison, a place, you will agree, hardly suited for meditation, and where I have been dumbfounded at my own listlessness in keeping my mouth shut at my stupid, "spiteful critics."[30]

Galois here lumps together the mathematicians of the academy with the government authorities, claiming that in practice there is no distinction between the two. Both together are responsible for his professional difficulties, as well as for his prison time. All are, for Galois, representatives of an implacably hostile and stupid "establishment" that has many incarnations but is ultimately faceless. For its representatives, Galois has nothing but scorn:

> I must tell you how manuscripts go astray in the portfolios of the members of the Institute, although I cannot in truth conceive of such carelessness on the part of those who already have the death of Abel on their consciences[31] . . . I sent my memoir on the theory of equations to the Academy in February of 1830 (in a less complete form in 1829) and it has been impossible to find them or get them back . . . Happy voyager, only my poor countenance saved me from the jaws of wolves.[32]

Galois, in his own eyes, is a meek lamb confronted with the wolves of the academy, who take pleasure in devouring promising young mathematicians. His rejection by the academy, he claims, guarantees that his work will not be taken seriously by the public, but will be received with an indulgent smile at best. All of which, he adds, "is to prove to you that it is knowingly that I expose myself to the laughter of fools."[33]

Here, in Galois' own hand, all the basic elements of what would become the "Galois legend' are already in evidence. The brilliant young man presents a work of genius to his superiors, innocently believing that they will recognize its value and welcome him into their ranks. Instead, he is confronted by a pack of wolves—stupid, incompetent, but fierce in their defense of their own privileged position. They cavalierly dismiss the upstart without ever giving his work serious consideration. In his disgust the young genius turns to the broader public for recognition, but even in this arena his reputation has been smeared. Most likely, the man rejected by the academy will be received with pity and condescension by those he knows to be his inferiors.

This is the fate of the pure man of genius, in Galois' eyes: to be cast out and marginalized by a cold and ruthless establishment and despised by the general public. His only refuge is his own purity and the beauty of disembodied mathematics itself. When his friend Auguste Chevalier suggested that he had been soiled by the "putrefied filth of a rotten world," Galois insisted on his own inviolable purity: "I am disenchanted with all," he responded in his letter of May 25, 1832, "even the love of glory. How can a world that I detest soil me?"[34] Four days later, on an empty Paris street, Galois took the ultimate step in his separation from the world he despised.

It is not impossible that the romantic Galois, who certainly had a flair for the dramatic, was also the author of the classic and concise summary of his life and career. An 1848 article on Galois in the *Magasin Pittoresque* reports that at the bottom of his letter to N.L. and V.D. written on his last night, Galois added a Latin postscript: "Nitens lux, horrenda procella, tenebris aeternis involuta" (a brilliant light, a fearful tempest, enveloped by eternal darkness).[35] I have found no corroboration that Galois indeed wrote these words, and they are quite possibly the invention of a later Galois acolyte intent on promoting the legend of his hero. Nevertheless, they beautifully capture Galois' view of his short life. When he crossed over to "eternal darkness" on the morning of May 31, 1832, it was the last and inevitable chapter of a tragedy that took place largely within his own heated imagination.

The story of the short life and tragic death of Galois soon became well known among the French literate public, for Galois, despite viewing himself as the perennial outsider, had many friends. Thousands attended his funeral, mostly young republicans like himself, eager to anoint him a martyr for the cause. Auguste Chevalier, the man to whom Galois' last letter had been addressed, published a heartrending obituary to his late friend in the *Revue Encyclopédique,* a journal that combined a scientific emphasis with republican politics. "Galois is famous for his ardent republicanism, for the judgments rendered against him, and for his long incarcerations; he will be far better known one day for his scientific genius," Chevalier promised less than four months after the death of his friend.[36] And so it proved to be.

In his obituary Chevalier detailed the long and sad tale of the young genius's encounters with the scientific establishment as Galois and his friends perceived them. In 1829 Galois' first memoir to the academy was lost by Cauchy, and a year later he submitted an essay for the academy's grand prize in mathematics. Galois, Chevalier wrote, "may not have suspected perhaps that the age of 18 years and the title of a student were altogether sufficient recommendations for making fun of his pretensions, and condemning without reading conceptions that are altogether new."[37] Galois, according to Chevalier, had hoped that his essay, if read carefully and conscientiously by the academy, would serve as compensation for the neglect of the previous year. Reality proved to be different: "This memoir, like the first, was lost," Chevalier wrote bitterly. "At that moment Galois resolved to abandon the way the academy offers a nascent talent to make itself known; painfully, he realized that this way was imaginary."[38]

According to Chevalier, it was the pain of his rejection by the academy and his anger at established institutions that stood in the way of true merit that led Galois to devote himself fiercely to the republican cause.[39] His long sojourns in prison, which resulted from his public defiance of the new regime of Louis-Philippe, proved disastrous for the young man. When he emerged from prison the first time, "his heart was swollen with despair; he was alone in the world."[40] The second time, "death awaited him upon his release."[41] Although Chevalier gave few specifics of the affair that led to the final duel, he left little doubt about who was ultimately responsible for the tragedy that befell his friend: it was the mathematicians of the academy, who turned away the gift of a young genius, condemning him to oblivion, despair, and ultimately death.

Chevalier concluded his tribute to his friend with a creed that may have been something of an anthem to republican revolutionaries in the 1830s. In the case of Galois, however, it took on special meaning: "The child of the poor, martyred by his genius, the heart compressed, hands tied together, the mind on fire, advances through life in fall after fall, or even torment after torment, to the morgue or to the scaffold."[42] The message is clear: Galois was the victim of a system that refused to recognize the extraordinary achievements of an outsider, a veritable "child of the poor." Overwhelmed by the fire of his genius that found no outlet, he was condemned to a life of torment and an early grave. Galois, according to Chevalier, was literally a martyr to his own genius.

A Martyr Resurrected

In the following years Galois' memory was kept alive by his prison friend Raspail, who in 1839 published his *Réforme pénitentiaire: Lettres sur les prisons de Paris,* with its heartrending recollections of Galois in prison.[43] Two years later the poet Gérard de Nerval, who also spent time with Galois in Sainte-Pélagie, published an essay titled "Mes prisons" that relates how Galois tearfully embraced him and bade him farewell upon Nerval's release.[44] Meanwhile, Alfred Galois and Auguste Chevalier never tired of trying to interest established mathematicians in the work of their departed brother and friend. In 1843 they succeeded.

Joseph Liouville (1809–1882) was 34 years old and a rising star in French mathematics in 1843, a professor at the École polytechnique and a member of the Academy of Sciences. He had never met Galois, but when he was approached by Alfred Galois with his brother's papers, he agreed to render an opinion on Évariste's published and unpublished work. Liouville was the first mainstream mathematician to give Galois' papers the careful and conscientious reading that their author had desperately sought years before, and he was enormously impressed. In 1843 he announced to the academy that Galois' papers contained important results, and in 1846 he edited the manuscripts and printed them in his newly founded *Journal de Mathématiques Pures et Appliqués* as the "Óeuvres Mathématiques d'Évariste Galois."[45] It was the breakthrough Galois' admirers had been waiting for, and it had an enormous effect on Galois' posthumous reputation. With Liouville's seal of approval, coming from a man at the heart of the French scientific establishment, Galois could no longer be dismissed as a hotheaded amateur treading the margins of sanity. From then on he was universally acknowledged as a true mathematical innovator.

Liouville's sober reflections in his introduction to Galois' "Oeuvres" stand in marked contrast to the emotional tone set by Chevalier and

Galois' friends. He criticizes Galois' excessive brevity when presenting his ideas and his difficult style, implying that his colleagues at the academy were not wholly to blame for missing the importance of Galois' discoveries. Liouville also suggests that the "illustrious geometers" of the academy, troubled by the obscure style of Galois' memoir, "were trying by the severity of their sage counsels to bring back to the right road a beginner of plain genius, though inexperienced." After all, he continued, "The author whom they censured was before them, ardent, active, he could profit from their advice."[46]

Liouville's conciliatory tone toward the academicians is understandable, considering that the chief target of derision for Galois' advocates, Augustin-Louis Cauchy, was still very much alive in 1846 and was Liouville's senior colleague. But it is also the case that his outsider's interpretation of Galois' troubled relationship with the mathematical establishment is probably closer to the truth than Galois' own impassioned view of things. In the end, however, Liouville does not hesitate to praise Galois' work in the highest possible terms. "My zeal" (in pouring over the Galois manuscripts) "was well compensated, and I experienced an ardent pleasure the moment when, after having filled out slight lacunae, I recognized the entire exactitude of the method."[47] Galois' method, he writes, "suffices in itself to secure our compatriot a place among the small number of 'savants' who have merited the title of inventors."[48]

The endorsement of Liouville marked a turning point in Galois' posthumous reputation. Over the next few decades his work was gradually absorbed into the mainstream of nineteenth-century mathematics, culminating with Camille Jordan's *Traité des substitutions* of 1870.[49] But just as significantly, the story of Galois' life moved from the radical margins of the French political spectrum to become a staple of the scientific establishment itself. The 1848 article in *Magasin Pittoresque,* just two years after Liouville's publication, still retains some characteristics of the radical fringe, where Galois' story was first disseminated.[50] Like Auguste Chevalier's obituary of 1832, it was likely authored by a friend of Galois, and the personal connection was enhanced by a sketch of the young mathematician, drawn from memory by his brother Alfred. Unlike later discussions of Galois, the article saw light in a popular magazine, not an official publication, and included Galois' unsubstantiated (and quite possibly mythical) postscript "Nitens lux, horrenda procella, tenebris aeternis involuta." All this is typical of the early reports on Galois, when his memory was preserved by his friends and radical republican comrades, but one can nevertheless sense a certain shift in tone. In place of Chevalier's bitter recriminations against the established mathematicians of the academy, the article mostly emphasizes Galois' rehabilitation by the academician Liouville.

Subsequent accounts of Galois are of a different character altogether, for they carry the official imprint of the French scientific establishment. Whereas Liouville was still concerned about maintaining working relations with some of the villains of the Galois legend, his successors in the following decades had no such concerns. Ironically, the legend of Galois, the young genius driven to an early death by a jaded scientific establishment, became a founding myth of the very institutions that had spurned him during his lifetime. Even the École normale, which had expelled Galois in disgrace in 1831, now embraced him as its brightest star.

In 1895 the École normale supérieure, celebrating its 100th anniversary, issued a collection of essays titled *Le centenaire de l'École normale*. For a piece on the influence of the school's most notorious student, the editors called on Norwegian mathematician Sophus Lie (1842–1899), whose work relied extensively on what was by then called "Galois theory." Lie obliged them with a nine-page article, "Influence de Galois sur le développement des mathématiques" (Influence of Galois on the Development of Mathematics), emphasizing the depth and importance of Galois' insights for the evolution of mathematics in the six decades since his death. Realizing, however, that Galois' fame rested not only on his mathematical genius but on his extraordinary biography as well, the editors decided that they needed a short summary of Galois' life to accompany Lie's essay. They turned to historian Paul Dupuy (1856–1948), who also happened to be the inspector general (*surveillant général*) of the École normale at the time. And so it came about that 65 years after the École's director, Guigniault, had clashed with Galois and expelled him from the school, the École's inspector, Dupuy, was charged with writing the young expellee's biography.

Dupuy took his assignment seriously. He assembled all the published materials on Galois and then combed the archives of the École normale, the Ministry of Public Instruction, the academy, and the hospital where Galois died. He searched the records of the city of Paris, Galois' hometown of Bourg-la-Reine, and the prison of Sainte-Pélagie, and acquired the complete transcripts of his trials. He then contacted Galois' surviving family and heard anecdotes and family traditions from them. He also acquired a sketch of Galois, drawn from life when he was 15 or 16 years old. By then it was clear to Dupuy that his project far exceeded the requirements and the space allotted to the piece intended to accompany Lie's mathematical essay. Instead, Dupuy decided to compose a full-fledged biography of Galois that would stand on its own. The resulting work, *La vie d'Évariste Galois,* was published by the École normale in 1896 as a separate issue in the journal *Annales scientifiques de l'École normale supérieure*.[51]

Dupuy's biography was, on the whole, balanced and sober. His research was as solid as it was comprehensive, and it provides the basis for all the accounts of Galois' life that followed. Nevertheless, there is no mistaking Dupuy's sympathies and his sense of kinship with the young mathematician, for Dupuy was a socialist and a leading "dreyfusard" who saw his own political battles mirrored in the struggles of radical republicans such as Galois over 60 years earlier.[52]

The source of Galois' troubles, Dupuy believed, was that he lived as a stranger among his fellow men, even among his fellow "patriots," as the republicans of the time referred to themselves. "His true fatherland, the most beautiful and the largest of all, is necessarily the one where so many noble intelligences, dispersed in all the corners of the world, fraternize among the rigorous conceptions and depths of mathematics."[53] Galois' true home, in other words, was a universe separate from our own unpredictable world: it was the pure and perfect world of mathematics, as rational as it is beautiful, to which only a select few ever gain access. Condemned to live his life in the crass world of men rather than in the Platonic heaven of pure mathematics, Galois was doomed.

"It is not rare," said Dupuy, "to hear mathematicians lament the brevity [of Galois' life]: for what would such a genius not give us, if death had not taken him at twenty! But no. Galois, it seems, had fulfilled his destiny."[54] If Galois had been admitted to the École polytechnique, as he had hoped, he would likely have perished on the barricades of July 1830. And if he had possessed sense enough to avoid the duel two years later, he would certainly have died in the disturbances of June 1832. Galois, according to Dupuy, was a stranger in our mundane world and was marked for an early death.

Around the time at which the École normale was embracing its errant student, other branches of French scientific officialdom were also hailing the genius and accomplishments of the man they had once rejected. In 1897, one year after the publication of Dupuy's biography, La Société Mathèmatique de France reprinted Galois' "Oeuvres," first published by

Liouville 50 years earlier.[55] This time the collection came with an introduction by the society's president, Émile Picard, a member of the Academy of Sciences and a leading mathematician of his generation, who wrote:

> In the presence of a life so short and so tormented, the admiration redoubles for the prodigal genius who had left such a profound mark on the sciences; the examples of precocious productions are not rare among the great geometers, but that of Galois is remarkable among all. Alas! It seems that the unfortunate young man sadly paid the ransom of his genius.[56]

The echo of Auguste Chevalier, who had written so many years earlier that Galois was "martyred by his genius," is unmistakable.

A decade later Jules Tannery (1848–1910), historian, teacher of mathematics, and professor of calculus at the École normale, published an updated collection of Galois' writings that included documents not available to the previous editors.[57] When in 1909 a plaque was placed at Galois' birth home in Bourg-la-Reine, Tannery was the main speaker. Galois, he explained, had dedicated his genius to exploring one of the most abstract fields of the most abstract of the sciences: mathematics. "It is an entire new world to study, an abstract, colorless, mysterious world. Those who know how to penetrate it, illuminate and direct it, are passionate about the faraway views that they have discovered."[58] Galois, said Tannery, was one of those blessed with the ability to penetrate this alternative world and bring us news of it. The notion that Galois belonged to this pure mathematical world rather than to our own material one, first suggested by his radical friend Chevalier 80 years earlier, had by now become the commonly accepted wisdom.

In his speech Tannery also took the opportunity to set the record straight on Galois' stormy career at the École normale:

> I owe to the position I occupy in the École normale the honor of saying this to you: I thank you, Mr. Mayor, for permitting me to make a full apology [*faire amende honorable*] to the genius of Galois, in the name of that school which he entered with regret, where he was misunderstood, from which he was driven out, and to which he is, despite everything, one of the most brilliant glories.[59]

Tannery's speech marked the culmination of a remarkable reversal of fortune in Galois' reputation in the French scientific establishment. Gone was the young man who possessed an irrepressible talent for making enemies of potential allies; nothing was heard of his "habit of impertinence," which provoked his radical comrades as much as distinguished academicians, and which allegedly drove his own mother out of her home. In their place appeared Galois as he himself and his closest friends had viewed him: an innocent who belonged in the world of abstract mathematical truth, not in the petty and conniving human world. His gift of genius coldly rejected, he was crushed by an uncomprehending and uncaring world. With this official makeover, Galois was installed in the pantheon of official French heroes.

From French Hero to Universal Icon

Even now the fame of Galois, the mathematical martyr, continued to spread. Over the next few decades this peculiarly French story, infused with the realities of Parisian institutions and political history, was transformed into a universal founding myth of mathematics. This was accomplished when Galois' story, until now accessible exclusively in French, was translated into English and presented to the far-larger English-speaking public. Two men in particular are responsible for the transformation of Galois from a national hero to a universal icon: George Sarton (1884–1956) and Eric Temple Bell (1883–1960).

Sarton was Belgian by birth but moved to the United States in 1915 during the German occupation of his homeland. Although trained as a mathematician, Sarton made his career in the history of science and is broadly considered the founder of the field as a professional discipline. Sarton was likely familiar with Galois' story from his student days in Ghent and was ideally positioned to bring this dramatic story before an American audience. In 1921 he published a short biography of Galois in the *Scientific Monthly*, which he reprinted repeatedly in subsequent years and eventually republished in his journal *Osiris* in 1937.[60]

There are no new facts in Sarton's biography of Galois, which was based closely on Dupuy's exhaustive research. But whereas Dupuy was intensely interested in the man Galois, seeking to rehabilitate him and bring him back into the fold of official French science, Sarton's Galois took on a different dimension altogether: for him, Galois was a universal archetype, the very model of a young scientific genius. "A life as short yet as full as the life of Galois is interesting not simply in itself but even more perhaps because of the light it throws on the nature of genius."[61] The young Frenchman, for

Sarton, was a saint and a martyr for science: "The soul of Galois will burn on through the ages, and be a perpetual flame of inspiration. His fame is incorruptible; indeed, the apotheosis will become more and more splendid with the gradual increase of human knowledge."[62]

Sarton went on to explain that Galois' short life made him an excellent case study of the differences between true genius and mere talent and dedication. Clearly the accomplishments of such a young man in so brief a time could not be considered the fruit of mere talent but must be the expression of a spark of true genius. "When genius evolves slowly, it may be hard to distinguish from talent—but when it explodes suddenly . . . we can but feel that we are in the sacred presence of something vastly superior to talent."[63] Unfortunately, according to Sarton, such an eruption of genius comes at a heavy price, and that is the most somber lesson we learn from Galois' short life: "If Galois had been simply a mathematician of considerable ability, his life would have been far less tragic, for he could have used his mathematical talent for his own advancement and happiness; instead of which, the furor of mathematics . . . possessed him and he had no alternative but absolute surrender to his destiny."[64]

The transformation of Évariste Galois into a universal icon was completed by Eric Temple Bell, whose short biography of Galois was by far the most popular chapter of his 1937 best seller, *Men of Mathematics.*[65] Bell was a mathematician at the California Institute of Technology who led a parallel life as a science-fiction writer, publishing his novels under the name John Taine. Bell's true aspiration, however, was to be recognized as a poet, and his classic work on mathematical biography showed signs of both his mathematical grounding and his literary ambitions.

Right from the start, *Men of Mathematics* was a huge publishing success. "There has recently appeared in this country a new kind of history of mathematics" written by "a poet who is known as a scientist, and a scientist who is known as a poet," raved historian of mathematics David Eugene Smith.[66] More than any other work, it has shaped the modern image of mathematics among the lay public, as well as among practicing mathematicians, and 70 years after its publication it is still in print and easily available. In colorful and dramatic prose Bell recounted the biographies of mathematicians from classical antiquity to modern times, often using their lives as the basis for broad speculations on moral and social issues. The life of the devout Blaise Pascal (1623–1662) was the occasion for Bell's musings on the dangers of religion to a great mind, and the story of the celebrated Bernoulli clan, whose members occupied many of the most prestigious mathematical positions in Europe in the seventeenth and eighteenth centuries, came under the appropriate heading "Nature or Nurture?" Not shy about giving full expression to his personal likes and dislikes, Bell titled his biography of Pierre-Simon Laplace "From Peasant to Snob," while William Rowan Hamilton (1805–1865) received more sympathetic treatment in "An Irish Tragedy." But no chapter in Bell's collection was more dramatic or more popular than his biography of Galois.

"Genius and Stupidity" was the title Bell assigned to his chapter on Galois, and he wasted no time in stating his thesis: "In all the history of science there is no completer example of the triumph of crass stupidity over untamable genius than is offered by the all too brief life of Évariste Galois," Bell announces in the very first paragraph of his biography.[67] In the rest of the essay he recounts the woeful tale of Galois' encounters with the supposed "fools" who crossed his path. First came his teachers at the Lycée Louis-le-Grand, who, according to Bell, were "good men and patient, but they were stupid."[68] Then, in quick succession, come the examiners of the École polytechnique, the director of Galois' "college" (Guigniault), the Paris secret police, and the authorities at Sainte-Pélagie. Bell does not go so far as to call the academicians Cauchy and Poisson fools, as indeed no mathematician would. Rather, he attributes their alleged hostility toward Galois to character flaws and institutional indifference. Overall, however, Bell insists that Galois' "powers were shattered before the massed stupidity aligned against him, and he beat his life out fighting one unconquerable fool after another."[69]

The profound impression left by Bell's biography of Galois on the minds of aspiring mathematicians and physicists is attested by leading lights of both fields: Leopold Infeld (1898–1968), James R. Newman (1907–1966), Fred Hoyle (1915–2001), Freeman Dyson (born 1923), and John Nash all claimed that Bell's biography of Galois had a profound influence on their lives and career choices.[70] Bell's version of the story of Galois was subsequently repeated in countless mathematics classes, related by teachers and professors to their students as a paradigm of the life of a mathematical genius. It became the "myth" of Galois, concerned with the universal archetype invoked by Sarton and Bell rather than with the historical man.[71] In a remarkable reversal of fortune, the marginal young man who was left to die alone on a Paris street came to embody the universal essence of a true mathematical life, his legend treasured by the mathematical community and known well beyond its confines.

Bell's angry tirade is very different in tone from Sarton's melancholy musings on the nature of genius, and he shows none of Dupuy's scholarly concern for historical fact and accuracy. His overall picture of the young mathematician's life, however, is not very different from the one painted by his predecessors. Galois, according to Bell, was simply too smart to fit in with the institutional settings in which he was placed. His genius soared so high above his contemporaries' limited intelligence that he could never bring himself down to their level. As a result, he was a misfit in a world dominated by mediocrities who, according to Bell, "were not fit to sharpen his pencils."[72] Bell's Galois lived on a higher plane of intelligence in a more beautiful and better world than the rest of us and was therefore a stranger among us. He was, simply put, too good for our flawed chaotic world, filled to the brim with narrow-mindedness and stupidity. Sarton, Dupuy, and probably Chevalier years before would all agree with Bell's assessment. And so would young Galois himself.

With Eric Temple Bell, the apotheosis of Évariste Galois had been finalized. The quarrelsome and unstable young man known to Sophie Germain and to his fellow radicals had completely disappeared from view. He was replaced by Galois the martyred innocent, the true genius who through his sufferings had become the patron saint of mathematics. His brief and troubled life had come to be seen not as a warning, but as the embodiment of the true mathematical life—wholly immersed in a world of beauty and perfection, forever a stranger in the midst of mere mortals. The cruel vicissitudes of earthly existence had claimed Galois' young life, but in the perfect world of mathematics Galois would live forever. "The truth of the work will survive," muses the fictional Galois of Tom Petsinis's novel *The French Mathematician* on the eve of his death, "while the chaos of my life will . . . Isn't that how it should be? Truth and beauty: In the end they are all that matter."[73]

4

THE EXQUISITE DANCE OF THE BLUE NYMPHS

Norwegian Prodigy

In 1826, when Galois was beginning his career of "impertinence" at the Lycée Louis-le-Grand, he was not the only young mathematician in Paris facing adversity. Another promising young man, a foreigner, was spending the year in Paris hoping to meet the leading mathematicians of the day and interest them in his groundbreaking work. Although he managed to meet many of the top academicians, his stay was not a success, and he was growing increasingly despondent over the polite indifference with which he was treated. After seven months he left the city and headed to Berlin and then home to Norway, having accomplished very little. His name was Niels Henrik Abel.

In personality and manners Abel could hardly have been more different from Galois. Whereas the Frenchman was a born rebel, irreverent and impertinent to friend and foe alike, Abel was a friendly and conventional young man, seeking to make a place for himself in middle-class society. And whereas Galois was a radical firebrand, Abel steered away from politics, devoting his creative energies exclusively to mathematics. It is quite remarkable, therefore, that in the annals of mathematics the two are viewed almost as identical twins, nearly interchangeable and sometimes confused with each other. This is not only because they worked on closely related problems, although their contributions to the theory of equations and what later became modern algebra are beautifully complementary. It is, rather, because their lives are viewed as similar in certain crucial ways and are taken to convey an identical moral lesson to subsequent generations of mathematicians.

Abel was nine years Galois' senior, having been born in 1802 on the island of Finöy in Norway and raised in the small town of Gjerstad, where his father Sören Georg Abel (1772–1820) was a Lutheran minister.[1] In 1815, at age 13, Abel left home to study at the Cathedral School in Christiania (modern Oslo). Although the school had a long and impressive tradition, it was then at a low ebb as many of its best instructors had moved to the new University of Christiania, which had opened its doors two years before. During his first two years at the school Abel's work was unremarkable at best. But in 1817 the school's mathematics teacher was fired for physically abusing a student. The teacher's replacement was 22-year-old Bernt Michael Holmboe (1795–1850), a graduate of the school and a former assistant to Norway's foremost scientist, the astronomer Christoffer Hansteen (1784–1873). Abel's career took a dramatic turn.

Holmboe was a competent mathematician and quickly recognized the outstanding mathematical talents of his student. With Holmboe's guidance Abel was soon reading the works of the great mathematicians of the past century—Newton, Euler, Laplace, Carl Friedrich Gauss, and Lagrange. It was not long before the student surpassed his teacher and ventured out on mathematical investigations of his own. The excited Holmboe, who realized that he had discovered true genius, soon introduced him to his own mentor, Hansteen, and to the university's professor of mathematics, Sören Rasmussen. As a result, even before he began his university studies, Abel was already becoming a member of the tight-knit Norwegian scientific elite.

Abel's contacts extended even further. While still a student at the Cathedral School, he set out to resolve a classic mathematical problem that had challenged mathematicians since the Renaissance: finding a standard formula for solving the quintic equation (an equation of the type $ax^5 + bx^4 + cx^3 + dx^2 + ex + f = 0$). Abel, who later proved that such a solution is impossible, was convinced for a time that he had solved the problem, and Hansteen duly forwarded Abel's paper to the Danish mathematician Ferdinand Degen (1766–1825). Rather than having Abel's piece published by the Danish academy, as Abel had hoped, Degen responded by asking for more details and the resolution of specific equations. Abel set immediately to work but succeeded only in locating the error in his initial solution. For his part, Degen was sufficiently impressed with the Norwegian prodigy to suggest to him that he concentrate his efforts on fields that were the focus of interest in the centers of mathematical learning.

Abel's personal life had meanwhile suffered a dramatic reversal. His father, Sören Abel, had been deeply engaged in the Norwegian nationalist movement, which was working to create government institutions separate from Sweden, Norway's suzerain. In 1818 Sören Abel was elected for the second time to the Storting, the Norwegian legislature, but returned home in disgrace within a year after making unfounded allegations of corruption against fellow representatives. When he died soon after, heartbroken and given to drink, he left his widow and children with only a small pension to live on. It was far from sufficient, and the young Niels Henrik spent the rest of his life trying to recover from the financial setback caused by the death of his father.

Abel entered the university in 1821, and for the next few years he was supported mostly by a small stipend paid personally by Hansteen and Rasmussen. In 1823 he published the first of a steady stream of articles that appeared in Hansteen's *Magazin for Naturvidenskaben.* Aided by a grant from Rasmussen, he traveled to Copenhagen to meet Degen in person, and there, while staying at the home of his uncle, he met his future fiancée, Christine "Crelly" Kemp. In the same year he returned to the problem of the stubborn quintic and this time succeeded in proving that such equations are not solvable in radicals. In other words, he showed that there is no standard algebraic formula for solving an equation of the fifth degree. Abel was so pleased with his success at solving a problem that had stumped generations of mathematicians that he had the proof printed and sent to the leading mathematicians of Europe. He received no response.

In 1825 Abel set out on a two-year journey to the mathematical centers of Europe. Armed with a grant from the Norwegian government and accompanied by four other promising young Norwegians, Abel planned to spend the bulk of his time in Paris, which in the early nineteenth century was the indisputable mathematical capital of the world. Abel, however, was loath to part with his friends, whose interests directed them toward Germany and Italy, and so found himself in Berlin in the fall of 1825. This unexpected detour proved fortuitous, for in Berlin Abel met August Leopold Crelle (1780–1855), the man who would make Abel's reputation. Crelle was an engineer by training but had an intense interest in higher mathematics and was well connected in international scientific circles. He and Abel forged a strong bond that year in Berlin that proved advantageous to both. That very year Crelle founded the *Journal für reine und angewandte Mathematik,* which became popularly known as "Crelle's journal" and remained the foremost German mathematics publication throughout the nineteenth century. The very first issue contained no less than seven articles by Abel, and the following issues included even more.

While Abel was in Berlin, Rasmussen, his former patron, resigned his position as professor of mathematics at the University of Christiania. The two leading candidates to replace him were Abel and Holmboe, who enjoyed the advantage of being on the spot when the position became available. Despite Abel's growing international reputation, the choice fell on Holmboe, who, as Abel himself could attest, was a superb teacher. Abel, despite his urgent financial needs, warmly congratulated his former mentor, and they remained close friends for the remainder of Abel's life. Where Galois had a way of making enemies out of friends and potential allies, Abel, to the contrary, remained a loyal friend even to his rivals.

Abel set out from Berlin in the early spring of 1826, but instead of heading directly to Paris, he accompanied his friends on an extended tour of Italy. He finally reached Paris in July, only to find that the university was out for the summer, and many of the people he had come to meet were away on vacation. Even when they returned, they did not seem eager to meet with the young visitor, and the reticent Abel was seemingly reluctant to press his case. His attempt to meet with Adrien-Marie Legendre (1752–1833), for example, ended with him introducing himself as the elderly academician was getting into his carriage, and that was that. Rather than making the rounds among leading mathematicians, Abel spent his time working on a memoir on a special class of periodic complex functions known as elliptic functions, which he and others would ultimately consider his most important work. On October 30 he presented it to the academy at a public meeting, in which Cauchy and Legendre were appointed as referees. Abel spent the rest of his time in Paris awaiting their verdict. It never came, and the memoir was not heard from again until after Abel's death.

Despite his disappointments, Abel did make some significant contacts during his stay in Paris. He spent time with the brilliant young German Johann Peter Gustav Lejeune Dirichlet (1805–1859), who was also in Paris making a name for himself in mathematical circles. He also made the acquaintance of Frédéric Saigey, the republican editor of the *Bulletin des sciences mathématiques,* the same scientific journal in which Galois would publish his work a few years later. Abel wrote several articles for the journal reviewing the recent publications in Crelle's journal, in which his own articles featured prominently. In January 1827, his financial resources exhausted, Abel left Paris and returned to Berlin, where Crelle welcomed him with open arms.

Abel remained in Berlin through the winter and early spring, during which time he composed his longest paper, "Recherches sur les fonctions elliptiques," which was published in Crelle's journal in two parts the following year. Crelle himself, meanwhile, was hatching his own plans for Abel's future: "Crelle has bombarded me mercilessly to get me to remain here," Abel wrote to his friend Christian Boeck, one of the young men with whom he had traveled from Norway. Indeed, Crelle was using all his influence in the Prussian government in an effort to secure an appointment for Abel in Berlin. In the meantime, as the negotiations dragged on at their normal bureaucratic pace, he tried to keep Abel in Berlin by offering him the editorship of his journal. All to no avail: "[Crelle] is a touch exasperated with me because I say no," Abel wrote to Boeck. "He does not understand what I shall do in Norway, which he seems to think is another Siberia."[2] In early May Abel started home from Berlin and, after passing through Copenhagen, arrived in Christiania on May 20, 1827.

Despite his impressive list of publications, Abel's situation in Christiania was difficult. He had no regular position, his request for a grant from the government was turned down, and his younger siblings still relied on him for support. For a while he lived only on a small stipend from the university, but things improved considerably when his friend and patron Hansteen set out on a two-year expedition to Siberia to investigate the earth's magnetic field, and Abel was named as his replacement. All the while, Abel's international reputation was growing by leaps and bounds. In early 1828 he learned that the young German mathematician Carl Gustav Jacob Jacobi (1804–1851) was also conducting research on elliptic functions and publishing results that were closely related to his own work. Suddenly aware that a rival was at hand and anxious to preserve priority and credit for his work, Abel quickly published a series of articles containing his major results in the field. These were to prove his last.

The rivalry between the two rising stars, Abel and Jacobi, caught the attention of the mathematical world. Legendre in particular, who had missed Abel in Paris a few years before, came to appreciate the importance of his work and entered into a correspondence with the young Norwegian. Aware that Abel's position in Christiania was tenuous, he joined three other academicians in writing a letter on his behalf to King Charles XIV of Sweden and Norway. They requested that the king support this "outstanding man" Abel and at the very least appoint him to the Royal Academy in Stockholm.[3] This letter produced no effect on Abel's employment situation, and there is no evidence that he even knew of it. Crelle, however, sensed the rising interest in Abel from other European scientific centers and redoubled his efforts to secure a position in Berlin for his friend. He now recruited the famed naturalist Alexander von Humboldt to the cause of bringing Abel to Berlin, and together they kept up steady pressure on the Prussian Department of Education.

But it was not to be. Abel spent Christmas of 1828 with the wealthy Smith family in the town of Froland, where his fiancée Crelly served as a governess. On January 9, when he was set to return to his teaching duties in Christiania, he fell violently ill, coughing and spitting blood. For the next 12 weeks Abel lay sick in Froland, cared for by Crelly and the Smith family. Diagnosed with "galloping consumption," or pulmonary tuberculosis, as it is known today, he died on April 6, 1829, at the age of 26. A letter from Crelle, announcing that his appointment in Berlin had finally been secured, arrived only a few days later. The following year the French Academy of Sciences awarded its grand prize in mathematics jointly to Abel and Jacobi, with Abel's portion of the prize going to his widowed mother.[4]

What are we to make of the short life of the young Norwegian? No doubt it was a difficult one. His father's early death and the financial hardships that followed, the pressures of supporting his siblings, and his inability to secure a regular academic position all contributed to a feeling of economic insecurity that is much in evidence in Abel's letters. "I am as poor as a church rat," he wrote to his confidante, Mrs. Hansteen, in the summer of 1828.[5] Add to that the chronic bad health that had dogged him since his second visit to Berlin and ultimately killed him, and one must conclude that Abel did confront more than his share of troubles during his 26 years.

Nevertheless, unlike Galois, Abel never considered himself the object of persecution or the victim of a faceless establishment. Quite to the contrary, while still a student at the Cathedral School, Abel was already welcomed into the homes of the leading scientists in Norway. During his difficult early days at the university it was they, and in particular Hansteen and Rasmussen, who were his chief sponsors and for a while supported him out of their own pockets. Being sent on a tour of Europe at government expense as one of Norway's "best and brightest" was certainly a great honor, even if it did not lead immediately to regular employment.

Although Abel's visit to Paris was disappointing, his two visits to Berlin were unquestionably triumphs. Abel's friendship with the influential Crelle, who appreciated the importance of his mathematical contributions, and his extensive publications in Crelle's journal laid the foundations for his rising international reputation. Finally, Crelle's unceasing efforts to bring him to Berlin pointed to a bright future in one of Europe's rising centers of mathematical research.

One could certainly take issue with the decision to appoint Holmboe rather than Abel to the vacant chair of mathematics at the University of Christiania. But in 1825, when Rasmussen resigned his post, Abel was still very much an unknown quantity, not the rising star of European mathematics that he would be three years later. Holmboe was seven years older, an accomplished teacher, and perfectly qualified for the position. He was also in Christiania when the decision was made, whereas Abel was abroad, a factor that inevitably influenced the selection. So although the university's choice certainly appears misguided in retrospect, things were not nearly so clear at the time. The choice of Holmboe over Abel was not an unreasonable one, and Abel, while disappointed, did not view it as a personal snub.

By 1828, when he was 26 years old, Abel was a bright star in the mathematical firmament. His competition with Jacobi was being followed closely by the leading mathematicians of Europe, and many, including Legendre, believed that Abel's work was superior to that of his rival. He was now corresponding regularly with Legendre, who along with Jacobi was expressing open admiration for his work. In an unprecedented move, four members of the Paris Academy appealed directly to the king of Sweden on his behalf, and in Berlin Alexander von Humboldt, perhaps the leading man of science in Europe, had joined the efforts to secure his appointment.[6] This is hardly the record of a man being persecuted by a hard-hearted establishment, and Abel, unlike Galois, never entertained such suspicions.

He was poor, of course, a condition that troubled him greatly, but by all accounts this was a temporary condition that would be alleviated as soon as he received his appointment in Berlin or elsewhere. He was also sick with tuberculosis, a disease that had no remedy and tragically ended his life. On his deathbed Abel was despondent over his misfortune and the injustice of dying when he felt that he had so much to contribute to the world. "I am struggling for my life!" he shouted repeatedly at his caretakers in moments of agitation.[7] But he never claimed that his illness was in any way connected to his endemic poverty, or that it stemmed from mistreatment by the scientific establishment. For him, early death was an unfathomable misfortune and an inexplicable act of God. The notion that it stemmed from his persecution by men simply never occurred to him. It is therefore remarkable to discover that young Abel's body had barely settled in its grave before accusations began flying from Berlin to Paris to Christiania, seeking to place blame on the true culprits in the young genius's death.

Who Killed Abel?

From the very moment of his death, the historical man Abel faded away and was replaced by a mythical being that would have been quite unrecognizable to the young mathematician himself. In life Abel was a sociable young man who easily charmed his way into the bosom of influential families such as the Hansteens and the Smiths. He was so sociable, in fact, that during his European trip he preferred to travel for months with his friends through Italy rather than turn to his own destination, Paris. Abel, unlike Galois, also possessed a reasonably practical mind-set and was quite capable of looking after his own interests. His early alliance with the leading men of science in Norway, his mutually beneficial friendship with Crelle, his voluminous and frequent publications, and his correspondence with Legendre and Jacobi in the last year of his life all show that Abel was no stranger to the invaluable art of self-promotion. But with Abel dead and buried, all this was instantly forgotten. In its place emerged an awkward young man who was never at home in the world and did not know how to act in society. This "Abel" sought refuge from the turbulent world of men and affairs and found it in the perfect and beautiful realm of mathematics. Only there was he able to find the peace and certainty he craved.

The tone was set early on in a commemorative poem by one of Abel's students, Hans Christian Hammer, published in May 1829 in the weekly *Den Norske Huusven*. Abel, wrote the admiring Hammer, through "the silvery blue ether, had fathomed the depths of the pale blue nymphs' exquisite dance."[8] Abel's genius, Hammer suggested, was his ability to see into a different and more beautiful world that was barred to the rest of us of us. This poetic imagining of Abel was soon followed by the "official" obituary, written by Holmboe, who had been Abel's teacher, friend, and ultimately rival. So great was the Norwegian public's fascination with the departed young genius that even before the obituary was published in the November 1829 issue of the *Magazinet for Naturvidenskaberne,* it was already printed and sold separately as *A Short Account of the Life and Work of Niels Henrik Abel.*[9]

Holmboe's essay was a disappointment to his expectant readers. Instead of waxing lyrical on the mysteries of genius plucked so soon from this world, Holmboe gave a matter-of-fact account of Abel's life and career. He recounted his friend's early years and their meeting at the Cathedral School, his tour of Europe, and the triumphs of the final years before his death. To lend support to Abel's importance as a mathematician, he quoted extensively from Legendre's letters to Abel, which contained great praise of the young Norwegian alongside technical mathematical discussions. Perhaps Holmboe remembered his friend as he was rather than as he would soon be imagined, or perhaps he felt the awkwardness of his position as the man who had been appointed to the post that was now viewed as having been Abel's by right. Be that as it may, Holmboe's account was judged accurate and informative but uninspiring. The growing legend of Abel was still searching for a spokesman.

It soon found him. Christian Peter Bianco Boeck was a student (later a professor) of veterinary medicine and had been one of Abel's traveling companions during his tour of Europe. He was also editor of the *Magazinet,* and to Holmboe's cold-blooded tribute to Abel in his publication he appended his own warm personal recollections. According to Boeck, although Abel appeared happy and sociable, this was not truly the case: "To those who did not know him intimately, his temperament may have appeared to be gay; those who saw him occasionally . . . may even have considered him frivolous." This, however, was a false impression: "On the contrary, his mind was serious and he felt deeply. Often he would be very melancholy, but he tried to hide it from most by forced gaiety and an indifferent attitude." Even in Berlin, where Abel was "extremely satisfied with his pleasant and useful association with Crelle," he still "often drifted into very dark moods."[10]

The reason for this, according to Boeck, was Abel's constant worry about his financial situation and professional prospects. "Only rarely did he seem to have a ray of hope that a position might be provided for him, but mostly no reasoning could encourage him." The specter of poverty and professional insecurity haunted young Abel, wrote Boeck, and his only refuge was in his work: "By clearing up some theorem he was working on he could, for a few moments, forget everything else; he was perfectly happy."[11] Years later, when he was one of Norway's leading scientists, Boeck elaborated on this picture with an anecdote from the days when the young Norwegian travelers shared a flat in Berlin. Every night, Boeck related, Abel would get out of bed, light a candle, and set to work on his mathematics. At one time Abel had resolved a difficult proof in his head but had forgotten the sequence of steps before he had a chance to write it down. For days he brooded, frustrated and unable to recover the lost proof. But one night, sitting at his desk, Abel woke all of his companions with a whoop of joy. The entire sequence of the proof had come back to him in a sudden flash of clarity.[12] Abel's disposition, according to Boeck, did not improve when he returned to Norway. Despite his more secure financial situation as Hansteen's replacement, and despite Crelle's encouraging missives from Berlin and his own growing fame abroad, Abel, according to Boeck, was despondent over his situation in Christiania. Lacking a regular position, he saw little prospect of obtaining one. This, Boeck wrote in the *Magazinet,* led directly to his death:

> He was compelled through his work to maintain and increase the esteem and fame already achieved, which promised to open brighter prospects for him. But his labor contributed to his early death. He sought little rest or diversion; excessive studies influenced his nerves; the sedentary life had a harmful influence on his chest—then another tiny thrust and he succumbed.[13]

Boeck spared no words in pointing a finger at the parties he considered guilty in Abel's demise:

> In his own country, he hardly found recognition for his knowledge and no encouragement. Abroad, the famous scientists praised the work of his genius, his discoveries, and his scientific merits; a foreign university, one of the best known in Europe, considered it an honor to count him among its teachers; a foreign government wished to provide for his welfare and give him an honored position.[14]

But at home he faced nothing but rejection and insecurity.

Boeck's depiction of Abel fits the familiar mold of the "tragic young mathematician" to the letter and set the tone for all future discussions of Abel's life and its meaning. The cheerful and playful Abel, who liked to romp in the snow with the Smith children, becomes a mere facade in Boeck's recollection. The refined and sociable young man disappears from view and is replaced by a morose and awkward young man, forever suffering in his interactions with others, misunderstood and mistreated by the scientific establishment of his own country. Only in the pure and beautiful realm of mathematics, to which he—almost alone—has access, is he fully at home and happy. Sadly, Abel was forced to live his life in a petty world in which he did not belong and that he did not understand, and the results were predictable. Dejected and not knowing where to turn, he soon lost his way and died, in effect, of a broken heart. Although one can hardly conceive of two people whose character and biography were more different than Galois and Abel, both were posthumously forced into the straitjacket of this standard biography. Oddly, the quarrelsome Galois and the amiable Abel became near biographical twins in the eyes of posterity.

Boeck's charge that Abel's colleagues in Christiania were somehow responsible for his early death touched a raw nerve at the university and sparked a controversy that lasted for years.[15] Hansteen was blamed for embarking on an expensive expedition while Abel was left penniless, and Holmboe for accepting a professorship that was now seen as Abel's by right.[16] It is ironic that it was Abel's closest friends and supporters in Norway, Hansteen and Holmboe, who were blamed the most for his death by those who knew little of Abel in his lifetime but saw themselves as his champions after his death.

Boeck's tribute to his departed traveling companion resonated well with a eulogy from Abel's staunchest mathematical advocate, August Leopold Crelle. In a tribute dated June 20, 1829, and published in his journal, Crelle published a long essay in which he dwelt on Abel's role in launching the publication and praised his genius and contributions to the science of mathematics:

> All the works of Mr. Abel carry the imprint of extraordinary, and sometimes truly amazing, ingenuity and force of mind . . . He penetrated often to the bottom of things with a force that seemed irresistible. He attacked the problems with such extraordinary energy, he beheld them from such height, and soared high above their actual state, that all difficulties seemed to vanish before the victorious onslaught of his genius.[17]

Abel's personal qualities, according to Crelle, fully matched his superlative abilities as a mathematician:

> It is not only Mr. Abel's great talents that render him so worthy of respect, and that make us always regret his passing. He was equally distinguished by the purity and nobility of his character, and by a rare modesty that rendered him as amiable as his genius was extraordinary. The jealousy of the merit of others was altogether unknown to him. He was far from that greed for money or titles, or even of pursuing renown . . . He perceived too well the value of the sublime truths that he sought, to put such a low price on them. He sought the compensation for his efforts in the results themselves. He rejoiced almost equally in new discoveries whether they were made by him or by another . . . He never did anything for himself, but all for his beloved science.

Crelle concluded the hagiography of his departed friend with musings about the unsuitability of such saints to our world:

> Perhaps such selflessness is misplaced in this world. Abel sacrificed his life for science without looking after its proper preservation . . . Give glory to the memory of this man, equally distinguished for his most extraordinary talents as for the purity of his character, one of those rare beings that nature rarely produces in a century![18]

Crelle's obituary is nothing less than a beatification of Abel. Not only did Abel's genius place him on a higher plane of existence than the rest of us, but his moral character was unquestionably Christlike: like Christ, he brought a precious gift to the world and was unwilling to sell it for personal gain; and just like Christ himself, he paid for his saintliness with his life. Both intellectually and morally, Crelle concludes, Abel belonged to a better and more perfect world. He brought us precious gifts from that higher place, but in our own base world he was a stranger, and all too soon he bade us farewell.

Crelle reiterated his views of Abel's life in 1832, when he wrote a preface to Jacobi's review of a work by Legendre based on Abel's results. Legendre gave full credit to the young Norwegian, writing that "it has been a great satisfaction to me to pay homage to Abel's genius by showing the value of the brilliant theorem he discovered; it might be called his 'monumentum aere perennius.' "[19] Crelle fully agreed: "This work acquires a peculiar value because it creates a worthy memorial to the genius of Abel, who was called away so regrettably in the far north . . . Unfortunately, it is very probable that he has taken with him to his grave precious treasures of new discoveries from the land of truth, mathematics."[20] This, according to Crelle, not our own imperfect world, was Abel's true home.

As it happened, Crelle was not the first to publish an obituary of Abel in a European scientific publication. Frédéric Saigey had been the scientific editor of the *Bulletin des sciences mathématiques* and had befriended Abel during his stay in Paris in 1826. In 1829 he joined forces with François-Vincent Raspail, Galois' future prison companion, to launch a new scientific journal, the *Annales des Sciences d'Observations*. The new journal lasted only four issues over two years, and a significant part of its function was to give vent to the editors' frustrations with the narrow-minded scientific establishment. To them, Abel's death was not just a personal tragedy but also a vivid example of what can happen when the leading scientific institution of the day is run like an exclusive club.

In an obituary published in May 1829, barely a month after Abel's death, Saigey reminisced about the time he had spent with Abel three years before. He told of Abel's poverty, his great learning, and his deep disappointment at his treatment by the academy.

> What I wanted to point out . . . is the fate which inexorably befalls a young man who presents himself to this high scientific court, with no other recommendations than his own works. It consists of a small group of scientists, men who have reached the age when thoughts are predisposed to turn towards the past, men who receive scientific innovation with anxiety and rancor.

These men, according to Saigey, were interested only in their own work, and if they paid attention at all to young scholars, it was only to direct them to continue the academicians' own line of research. Such advice should be flatly rejected, Saigey concluded: "Young scientists, do not listen to anything but your own inner voice, which tells the tasks best suited for your own inclinations and abilities . . . The device must be: Objectivity toward the facts and freedom in the choice of views."[21]

For Saigey, Abel's story exemplified what happens to true talent when faced with a fossilized scientific institution like the academy, interested only in perpetuating its own existence. Crelle had emphasized Abel's sanctity of character and his unsuitability for mundane life, whereas Saigey was more interested in the hostility of the scientific powers that be. But while Crelle and Saigey differed in emphasis, both adhered closely to the familiar tale of the unacknowledged mathematical genius. Abel, Crelle argued, was a visitor from the perfect and beautiful "land of mathematics" who was ill suited to promoting his own interests in our petty world. Conversely, Saigey pointed out, our worldly institutions were themselves too flawed to acknowledge and accept the gifts of young genius. From there it was only a short way to the inevitable end—disillusion, despondency, and early death.

Crelle's idealized depiction of Abel was perhaps to be expected, coming as it did in the months immediately following Abel's death. Saigey's protest against the academy's scientific monopoly was also not unexpected, as it formed a salvo in a constant barrage of criticism against the scientific establishment coming from Raspail and his circle. More surprising is the reaction of Guillaume Libri, the young Italian mathematician who now assumed the role of Abel's champion in Paris, the city that had given him the cold shoulder years before. Libri was the very man who was treated to Galois' "insolence," according to Sophie Germaine. Typically, Galois had offended a man who might have been a natural ally, for Libri was well known as a republican and radical activist. Born in Florence as Guglielmo Icilio Bruto Timoleone Libri, Count of Carucci dalla Sommaja, Libri was only 20 years old when he was appointed professor of mathematical physics at the University of Pisa in 1823. Within a year, while retaining his academic rank and salary, he moved to Paris and introduced himself to the leading mathematicians of the day. His urbanity and sophistication, and perhaps also his high birth, earned him a warm reception that was not afforded to the likes of Galois and Abel. In 1830, inspired by the revolution in Paris, he returned briefly to Florence and took part in a failed republican coup. He returned to Paris a revolutionary hero and an exile, and by 1833 he had been elected to Legendre's seat in the Academy of Sciences and appointed professor at the Collège de France.

This was a high point in Libri's career, but it did not end there. In 1840 Libri was appointed secretary to a royal commission charged with registering and cataloguing the holdings of French provincial libraries. It was soon noticed that precious volumes regularly disappeared from the libraries following Libri's inspections. When he subsequently held auctions of antique books of unclear provenance, an investigation was launched. For several years Libri's many powerful friends protected him from prosecution for stealing books from libraries under his supervision, but when a new regime came to power after the revolution of 1848, Libri was indicted. He managed to escape to England by the skin of his teeth, but accompanied by 30,000 antique volumes that he again auctioned off at a good profit. His personal charm proved as effective in London as in Paris, and his many English friends refused to believe the accusations against him. Libri finally returned to Florence shortly before his death in 1869, where he lived out his remaining days under a cloud of suspicion, but nonetheless a very wealthy man. A joint British and French investigation after his death established beyond doubt that Count Libri, suave and charming society man and talented mathematician was, first and foremost, a thief on the grandest scale.

All this, however, lay in the future. In 1833 the dashing young Italian was at the height of his success when he became the self-proclaimed champion of Abel in France. Unlike Saigey, there is no indication that Libri had met Abel during his stay in Paris in 1826, or that he showed any interest in him or his work in the following years. Certainly there is nothing in his biography of Abel that suggests a close familiarity with his subject matter.[22] Libri, however, was undaunted as he proceeded to fill in any lacunae in his knowledge of Abel with his own imaginings. Libri's Abel was possessed of "unusual modesty" and a "natural timidity" that prevented him from reaching out to leading mathematicians and making a name for himself. During Abel's visit to Paris "no one suspected the genius of the young man whose death two years later reverberated mournfully all over Europe." Upon his return to Christiania, Abel "could find no position, no help," and "had to seek refuge with his poor mother."[23]

"It was not so much poverty that burdened Abel, for men of Abel's character have loftier aims than money," wrote Libri, echoing Crelle's obituary. "It was rather his consciousness of his own superiority, without finding anyone who could comprehend the power of his genius." In the end, "his heart thus discouraged, an immoderate amount of work and sorrow destroyed his constitution." Even then, however, his love of science and mathematics was unconquerable: "It was under that state of lonesomeness and suffering that he composed the beautiful works that have been a source of admiration among the geometers."[24] All the familiar elements of the legend are in place here: the young genius is too modest and noble to make his way in the world. Conscious of his superior abilities, he presents his work to the mathematical community but is met with cold disinterest. Discouraged and overworked, he soon succumbs to illness and early death.

Having transformed Abel's life into this standard formula, Libri then followed Saigey in denouncing the established mathematicians who, he claimed, contributed to Abel's death. Unlike Saigey, the perennial outsider, Libri was writing from the very bosom of the French scientific establishment. This did not cause him to temper his attacks. He railed against the treatment of the memoir Abel had submitted to the academy in 1826, which had been lost and remained unpublished at that time. It was "through the nonchalance of the modern mathematicians," Libri wrote, that "Abel's memoir long remained buried in the papers of committee members."[25] He ended his account of Abel's life with a roar: "Here we raise our voice to demand an account of all those egotistical men who, by their indifference, have contributed to the shortening of Abel's life; we demand a reckoning of all the discoveries of which his death has deprived us."[26]

The tale of the mathematical martyr, it seems, requires its villains, although these can change depending on circumstances. For Boeck, in Norway, the chief culprits in Abel's death were the faculty of Christiania University, who did not offer Abel a position. For Libri and Saigey, in Paris, the ones most to blame were again local—the leading mathematicians and members of the Paris Academy of Sciences who ignored Abel during his visit in 1826 and lost his memoir. But the fact that different actors in different places are accused of doing pretty much the same thing only emphasizes the fact that the underlying story is exactly the same. It is the familiar tale of the young genius lost in a world of petty men and driven to his death by an uncaring establishment. The details and names in this basic outline can then be filled in according to circumstances.

Libri's involvement with Abel's legacy offers one final irony. With his biography of Abel, Libri had established himself as Abel's champion in the French capital. It was therefore quite natural that a few years later, when the Paris Academy finally took steps to publish Abel's lost 1826 memoir, it was Libri who was charged with locating the manuscript and bringing it to print. He did so successfully, and in 1841 the missing memoir was finally published by the academy and made available to the mathematical community. Sadly, the published text was based on a copy, as the original had by this time been lost once more, and its fate remained a mystery for over a century. In 1952 it was unexpectedly located by the Norwegian mathematician Viggo Brun in the Biblioteca Moreniana in Florence, where Libri apparently deposited it shortly before his death in 1869. Libri, who thundered against those who had ignored Abel's memoir, could not himself be accused of indifference to Abel's work, for he knew the value of the original manuscript all too well. He therefore stole it.[27]

Whereas Libri was using Abel's tragedy to increase his worldly possessions, as well as his public stature, others were using the legend of the young Norwegian to advance political and ideological ends. In 1829 Saigey had already claimed that Abel was a victim of an ossified and arrogant scientific establishment. As late as 1870 his friend François-Vincent Raspail rose in the French Chamber of Deputies to make the same point. This time, however, Abel was the victim not only of narrow-minded self-interest but also of class struggle. "It was about 40 years ago," recalled Raspail,

> at a time when the rapaciousness of the members of the Institute and the Academy of Science was at its height, when a young Swede presented himself to one of my friends.[28] My friend read the memoir which the young man intended to present to the Institute, and it seemed to him that it showed signs of a remarkable fantasy. He said to him: My friend, give your memoir to Monsieur Fourier in the Institute, to no one else; go to him personally, he will read it and fix a day. But he [Abel] presented his memoir to Cauchy, who at that time had accumulated positions yielding him as much as 50,000 francs a year, almost as much as the zoologist Cuvier, who managed to make 60,000 francs, while many young men were dying from hunger.
>
> The young man then went to Fourier, who received him in his small apartment the way he received everyone who worked in the sciences; he treated them with great consideration and never avoided a discussion of the problems which occupied them. Cauchy, the capitalist, on the other hand, to whom the young man had entrusted a copy of his article, managed to mislay it among his papers; Poisson, who had also received a copy, lost it in a similar manner.
>
> The young man was well received by my friend, who was an expert in this field. He was himself a man who was compelled to work for his living in the same way as he does today, poor and deserted. [Many members: his name, what is his name?]
>
> His name is Monsieur Saigey . . . Saigey offered the young man money, he invited him to eat at his home; but the young Swede blushed with shame to be supported in this manner at the expense

of a friend. One day he told that he was preparing to return to his home country, Sweden, but he went on foot!

He had left his memoir with Legendre, of whom you have all heard. Legendre had stated: See what a young man can pretend; he believes he has found the solution to a problem which I myself have worked on in vain for 40 years. Then he threw the paper on his desk.

But when he [Abel] traveled through Berlin, he left his work with a scientific journal edited by a scientist who received him with more kindness and understanding. Legendre was a decent man, and as soon as he saw the printed article, his conscience began to bother him. When he arrived home he sought the manuscript and then exclaimed to himself: actually, he has found what I was seeking, he has made the most difficult discovery in the world, he has produced the solution I have sought for 40 years.

Legendre grasped his pen, shaking with sorrow that justice must come so late, and wrote to the minister of education in Sweden, told him what a great scientist he possessed within his realm, and requested that he take him under his care, for Legendre had just been informed that the young man was starving. This was the reply: the young scientist was dead, dead from hunger.

Do you know who this man was? It was the same Abel whose name today is mentioned everywhere with the greatest admiration, because the whole world knows his memoir. He died when he was 25 years old.

Here one can see the consequences of the money hoarding among the members of the Institute. If they had not been allowed to accumulate money in this manner, they would have acted as my friend did, they would have taken care of this young man who showed the most eminent, yes, the most brilliant ability.[29]

There are many fanciful embellishments in this account, including the notion that Abel walked home from Paris, and that he died of hunger. The personal attacks, however, are not coincidental but are guided by the political affiliations of the people involved. Cauchy was an ultraconservative Catholic (an *ultra* in the parlance of the times) and therefore the archenemy of Raspail's radical republicans. One is therefore not surprised to find him accused of avarice and selfishness, traits that Raspail considered natural to followers of his creed. Fourier, on the other hand, was a child of the revolution, a high official under Napoleon, and a dedicated republican throughout his life. To the sympathetic Raspail, he was therefore an unassuming man of the people, whose "small apartment" was always open to aspiring young men of science.

Quite unexpectedly, in Raspail's account the tragic story of the young Norwegian mathematician turns into a case study of the oppression of the poor, hardworking masses by selfish, money-grubbing capitalists. We have already seen such claims made for Galois, who was described as a "child of the poor, martyred by his genius" in the obituary by his friend Chevalier. Perhaps there is indeed some justice in referring to Galois in these terms, for despite his bourgeois roots he did adopt the cause of social justice for the poor as his own. Abel, however, was quite apolitical and strictly middle class and would no doubt have been astonished to find his name used as a rallying cry for socialist revolutionaries. Nevertheless, the basic narrative behind Raspail's political allegory is very familiar by now: as always, Abel is the innocent genius who is crushed by the self-interested indifference of leading mathematicians. It is essentially the same story that was presented by Boeck, Crelle, Saigey, and Libri, and it is practically identical to the story that was also growing at the same time around Galois. With little regard for biographical specifics or circumstances, the legend of the tragic young genius was becoming a commonplace image of a mathematician.

A Mathematical Saint

Abel's formal canonization took place in 1902 at a conference held at the University of Christiania to mark the centennial of his birth.[30] It was a solemn affair attended by the king and the prime minister of Norway and the leading scientific lights of all of Europe. Among those who received an honorary doctorate from the university during the events were the mathematical and scientific giants of the age—Georg Cantor (1845–1918), Richard Dedekind (1831–1916), Ludwig Boltzmann (1844–1906), Camille Jordan (1838–1922), Felix Klein (1849–1925), David Hilbert (1862–1943), Andrei Markov (1856–1922), Henri Poincaré (1854–1912), Émile Picard (1856–1941), Lord Rayleigh (1842–1919), and Lord Kelvin (1824–1907). The mathematician Ludwig Sylow (1832–1918), who along with Sophus Lie had edited and published Abel's collected works, and who was considered Abel's mathematical heir in Norway, gave a welcoming address to the gathered dignitaries.[31]

Abel, Sylow proclaimed, played an essential role in the birth of a new type of mathematics. In the early nineteenth century, he explained, applied mathematics had already achieved great triumphs, especially in the fields of astronomy and physics.

> But just at the same time mathematics . . . started to turn its gaze back to the pure and abstract theories. [Gauss and Caucy] initiated thereby that great movement, which has run through the whole of the previous century, and which has reformed higher mathematics from its foundations at the same time it has enriched it with new theories . . . It was in this movement that Niels Abel took such a significant part that he will forever be counted as one of the greatest mathematicians ever.[32]

Abel, according to Sylow, played a crucial role in turning mathematics away from specific physical questions and toward a purely mathematical world. Much the same, we shall see, could be said of Galois and other mathematicians who fit the romantic tragic mold.

Sylow went on to emphasize Abel's amiability and nobility of character and closely followed Crelle in commenting that "such a great modesty does maybe not fit this world" and that "it may also have been seen as a weakness." "But notwithstanding the fact that during his whole life he had to fight against desperate economical circumstances on the one hand, and on the other to suffer from a lack of understanding of his worth here at home, he walked straight ahead on his road, without letting himself be discouraged."[33] Compared with the fierce denunciations of the academic establishment in Boeck's and Libri's essays decades earlier, Sylow's lament of the lack of appreciation of Abel's work in Norway sounds mild. This is perhaps understandable in a conference organized by the very university that failed to offer him a position, and on the occasion intended to elevate Abel into the Norwegian national pantheon. Nevertheless, the familiar outlines of Abel's tragic tale are recognizable in Sylow's speech: Abel, the young genius, was too pure for this world. Misunderstood and rebuffed by his colleagues, he paid a heavy price and perished before his time.

Distinguished visitors at the centennial also paid homage to Abel in similar terms. Andrew Russell Forsyth, who was then Sadlerian Professor of Pure Mathematics at Cambridge University, spoke "on behalf of those delegates for whom English is the home language."[34] The delegates "resemble pilgrims of old, on their way to a sacred shrine," he said, giving voice to the feeling that the centennial celebration was intended to bestow on Abel a scientific sainthood. Like religious martyrs of old, Abel was characterized by "his profound and penetrating insight, . . . his high courage, his unflinching belief in the loftiness of the work that he could do in life." And like a true saint, he was in touch with realms that were barred to the rest of us, having devoted "his mind and the full activity of his intellect to the exploration of unknown regions of analysis."[35]

Mathematician Émile Picard spoke for the Francophone delegates, joining Sylow in praising Abel for his leading part in the "revolution" that made mathematics "so precise in its fundamental concepts, and so inflexible in the logical rigor of its deductions." He went on to compare Abel's career, "so short and so tormented," with that of Picard's own countryman, Galois, "who also left behind a glorious legacy." "Abel and Galois," Picard lamented, "what reproaches these two names suggest to us! . . . Perhaps it is better suited in the meantime for geniuses of this order to disappear while still young, leaving behind them a sparkling wake."[36] The young Frenchman and the young Norwegian were, for Picard, mathematical and biographical siblings. Both died young and unacknowledged, and their memories are forever a reproach to us for failing to recognize their genius while there was still time to do so.

As befitting an elevation to sainthood, Abel's centennial was set to music—in this case, a cantata titled "Niels Henrik Abel" by Bjørnstjerne Bjørnson, with music by Christian Sinding. Verse after verse heaps praises on Abel's "mind," which "did streak towards the highest peak" under Norway's "great star." Numbers, Abel's object of study, are "steadfast as time—purer than snow, finer than air, yet stronger than the world." As a result, death can take no real hold of Abel's great mind:

> When he sensed his fate,
> Death came him to take,
> he asked death to wait.
> He pursued his design
> till he put the sign,
> the last
> under that which no one yet did know,
> nor comprehend,
> Now research can seek its end.

There are likely echoes of Galois' last night here in the suggestion that Abel held off his death until he could complete the "last sign" of his mathematical work before bequeathing it to posterity. There may also be echoes of Archimedes, who during the sack of Syracuse reputedly asked a Roman soldier to wait until he resolved a geometrical problem, whereupon the impatient soldier killed him. But there is also more in these stanzas, befitting the quasi-religious tone of the centennial proceedings. Although Abel sacrificed his earthly body to his mathematical work, his true self—his mind—lives on to eternity:

> Of his birth the star
> leads on to his cradle
> Wise men come from afar.
> Century?
> Seems just like yesterday.[37]

Abel is here a mathematical Christ. The magi of the world were summoned to the site of his cradle by a star to commemorate his birth, but even more so the eternal life of his genius.

In the 1902 centennial the legend of Abel reached its final and complete form. Abel was a mathematical saint who came to us from a pure mathematical universe and was ill fitted to care for himself in our own far-from-perfect world. Poor and unacknowledged, he turned his back on us and returned to his true home—the land of mathematics. Although we proved unworthy of his gift during his lifetime, his legacy will remain with us, forever brightening our lives with the brilliance of eternal beauty.

More than 100 years after the centennial celebrations, Abel's legend retains its force and vitality. Like Galois, Abel earned a chapter in E. T. Bell's *Men of Mathematics* of 1937, where his travails are described much along the lines set at the centennial. Titled "Genius and Poverty," the chapter describes how the genius Abel was done in by the callousness and incomprehension of established mathematicians in France and university professors in Norway. But perhaps the best indication of the perseverance of the Abel legend can be found in Arild Stubhaug's 1996 biography of Abel, the most recent and comprehensive available. Although Stubhaug was mostly known as a poet in his native Norway, the biography is very thoroughly researched and documented, as well as factual and sober in tone. In some places, however, the exalted tone of the 1902 centennial shines through, as when Stubhaug cites Abel's older contemporary, the romantic German poet Novalis, on the subject of mathematics. "All divine emissaries must be mathematicians," Novalis wrote; "pure mathematics is religion." Ultimately it is the title of Stubhaug's book that gives away the author's belief that Abel belonged to a brilliant and alien realm that lured him away from his mundane existence: *Niels Henrik Abel and His Times: Called Too Soon by Flames Afar.*[38]

The myths of Galois and Abel originated at the very time of their deaths. Galois, who railed against his mistreatment by the mathematicians of the academy, can to some extent be considered the author of his own story, although undoubtedly his final night and the duel added a new dimension to his legend. Chevalier in his obituary shortly after his friend's death already presented the story of the misunderstood young mathematician more or less in its final form, and in later generations this view of Galois was adopted by the scientific establishment itself. The more conventional Abel, although fully aware of his mathematical talents, never saw himself as a martyr, but after his death it did not take long for others to present him in precisely this light. Boeck, Crelle, and especially Libri combined to create a fictional Abel who was, in effect, "too good" for this world. The 1902 centennial with its religious overtones enshrined Abel as a veritable patron saint of mathematical learning, a status that he retains to this day.[39]

The similarities between the struggles of the two young mathematicians were first noted by Galois himself, who clearly had a grander notion of his place in history than the unassuming Abel. "I must tell you how manuscripts go astray in the portfolios of the members of the Institute," he wrote in his angry "Préface," composed in Sainte-Pélagie; "I cannot in truth conceive of such carelessness on the part of those who already have the death of Abel on their consciences."[40] Nevertheless, the now-popular notion that the two led nearly parallel lives did not take root until the turn of the twentieth century, when Galois was fully embraced by the mathematical establishment that had rejected him, and Abel was canonized in his centennial celebrations.

The delay itself is hardly surprising, given the sharp difference between the two in temperament and in the course of their careers, not to mention the manner of their deaths. The provocateur Galois who died in a pointless duel and the amiable Abel who died of an incurable malady could hardly be more different. Nevertheless, as we have seen, to Émile Picard at the 1902 Abel centennial the parallels between Galois and Abel appeared inescapable. This was also the case for Bell, who began his biography of Galois in *Men of Mathematics* with the words "Abel was done to death by poverty, Galois by stupidity."[41] The tradition continues in the 1957 biography of Abel by Oystein Ore, which includes a lengthy discussion of Galois' life and work, as does Stubhaug's biography of 1996. More recently, Mario Livio's popular account of group theory, *The Equation That Couldn't be Solved,* contains successive and parallel chapters on "The Poverty-Stricken Mathematician" (Abel) and "The Romantic Mathematician" (Galois).[42] The two have become inseparable from each other, and the telling of the story of one almost inevitably brings up the other.

This is not because the two young mathematicians truly bear any deep similarity to each other. The differences between their lives and careers were profound, and their personalities were as opposite as it is possible to imagine. The reason that they have become the conjoined twins of the history of mathematics is that their biographies have both been molded by the same romantic and fictionalized story. This story has shaped their memory almost from the day they died, and it is just as potent today as it was then. It is the romantic tale of the young and innocent genius whose gift to the world goes unacknowledged by his contemporaries, leading him to despair and early death.

This standard story tells us little about the actual lives and character of Galois and Abel, but it tells us a great deal about how mathematicians and the science of mathematics have been perceived since that age of High Romanticism. Mathematics was seen as a world unto itself, ruled by logic and beauty, separate from our own imperfect world and largely inaccessible. Mathematicians were those privileged with a glimpse of that world of perfection, who then tried to communicate its marvels to the rest of us who stayed behind. Is it a wonder that these individuals find it difficult to communicate their discoveries to those who spend their entire existence in our crass world? Hardly. Like the prisoners in Plato's cave metaphor, who refuse to acknowledge the truth presented to them by those who have seen the light, we also are bound to reject the gifts of the mathematical genius. Is it a wonder that a genius who has glimpsed a world of perfection would never truly belong in the world of ordinary men and women? Not at all. The stage is thus set for the tragedy of the young genius.

5

A MARTYR TO CONTEMPT

Worldly Men and Tragic Loners

Let us step back here and consider the vast gulf that separates the tales of the grands géomètres of the eighteenth century from those of their successors in the nineteenth century. On one side of the divide we have d'Alembert and Euler, "natural men" in tune with their surroundings, successful men of affairs who were the bright stars of their era and lived to a ripe old age. On the other side we have Galois and Abel, alienated loners who lived on the margins of society and died young, disillusioned, and unacknowledged by their peers. In the early decades of the nineteenth century, it appears, the image and persona of the practicing mathematician underwent a radical and startling change.

This, of course, is not to say that all Enlightenment geometers were successful and admired, like d'Alembert and Euler, or that all early nineteenth-century mathematicians were persecuted misfits who died young. Personal biographies diverge, and there were wrongly unacknowledged geometers in the eighteenth century, just as there were nineteenth-century mathematicians who were successful and famous men. But the significance of the tragic story of Galois and Abel, just like the story of d'Alembert in the previous century, was not that it correctly represented the lives of all mathematicians. For that matter, the legends were not even very good representations of the lives of their protagonists. Their significance lay in the fact that they became founding myths of the field, universal tales that captured the essence of what it meant to be a mathematician. And whereas the guiding legend of Enlightenment mathematics was that of Jean le Rond d'Alembert, Rousseauian "natural man," the corresponding tale for the nineteenth century was the story of Évariste Galois, alienated loner and mathematical martyr.

Marginal loner versus cultural hero, "martyr for truth" versus "natural man," Galois and Abel versus d'Alembert and Euler: such were the contrasts between the iconic images of mathematicians in the eighteenth and nineteenth centuries. A deep chasm appeared around the year 1800, separating Enlightenment and Romantic notions of what it meant to be a mathematician. As we shall see, these contrasting views were not just a matter of changing public images, or even just shifting ideals shared by the mathematicians themselves. They were, rather, fundamental shifts in the way the field of mathematics was perceived, its meaning, its rules, and its practice. The mathematics practiced by the tragic loners of the nineteenth century was profoundly different from the field as it was practiced by the grands géomètres of the previous age. The changing stories and legends about mathematicians and the mathematical life went hand in hand with changing technical approaches and practices in the field of abstract mathematics.

The legends of Galois and Abel required a villain, a person to serve as the embodiment of a narrow-minded establishment that would shunt aside the brilliant discoveries of the young genius. In both cases this role was filled by the most prominent and influential mathematician of their time, the man who towered over all his contemporaries in both accomplishments and honors—Augustin-Louis Cauchy. It is therefore startling to find that Cauchy would not have recognized himself in the image of the all-powerful establishment Academician that has come down to us. With good reason Cauchy considered himself a persecuted loner whose tale of woe and suffering at the hands of well-connected colleagues was on a par with the tragic tales of Galois and Abel. Astonishingly, the great Cauchy saw himself—and was viewed by his admirers—in much the same way as Galois viewed himself: a lone genius persecuted and marginalized by a jaded political and scientific establishment. Galois may have viewed Cauchy as his polar opposite and chief tormentor, but if one accepted Cauchy's view of things, the two were near twins.

The Conservative Mathematician

In many ways Cauchy's life and career contrast sharply with those of Galois and Abel. Cauchy was a child of privilege and affluence, whereas Galois and Abel, despite their middle-class background, lived in poverty. Cauchy did not die young but lived to the respectable age of 67 and was far from being unacknowledged by his peers. He was a lifelong member of the Paris Academy of Sciences and over the years amassed a long list of additional positions and honors, all testifying to his stature as the leading mathematician in France. His personal reputation, from his day to ours, is hardly that of a saintly victim but rather of a rigid ideologue, greedy in acquiring honors for himself but grudging in his praise and encouragement of others. We have already heard Galois' and Abel's views on Cauchy, whom they blamed for ignoring their work and losing their memoirs, as well as Raspail's denunciation of Cauchy in the Chamber of Deputies years later. To their voices we may add the recollections of another famous mathematician, Jean-Victor Poncelet, who in 1864 bitterly recalled Cauchy's cavalier and dismissive response when he had tried to present his work to him more than four decades earlier.[1] For them, Cauchy was the very embodiment of a hard-hearted scientific establishment that had no place for new men and new ideas.

It is worth noting, however, that Cauchy was no more popular within this supposedly monolithic establishment than he was among the outsiders. Although widely respected for his mathematical prowess, Cauchy was also universally disliked by his colleagues, having managed to turn many former and potential friends into enemies. Part of the disdain for Cauchy in official circles was undoubtedly political, for Cauchy was an *ultra*—a legitimist monarchist, advocate of the Bourbon Restoration and the Catholic Church.[2] Most of his colleagues at the academy, in contrast, were liberal republicans, secularists intent on carrying on the legacy of the Revolution into the new century. But much of the animosity was also undoubtedly personal, stemming from Cauchy's inflexibility and unwillingness to go along with the compromises and social niceties that greased the wheels of an institution such as the academy. All too often, for Cauchy, it was his way or nothing, and he did not hesitate to use whatever power was at his disposal to advance his cause.

To Cauchy, the notion that he was the embodiment of a powerful establishment would have seemed absurd. He was, in his own eyes, the exact opposite: a lonely bearer of truth, persecuted and marginalized by lesser men who refused to acknowledge the worthiness of his convictions. Unbending in his devotion to the truth, Cauchy marched on through thick and thin, never conceding an inch to his petty critics and willingly suffering the consequences of their hostility. In this sense Cauchy's view of himself was not unlike Galois' self-image or the legend of Galois that his friends promoted after his death. And although no tragic legend ever attached itself to the name of Cauchy, there is no denying that in the eyes of his conservative friends Cauchy was no less a mathematical martyr than Galois and Abel.

Augustin-Louis Cauchy was born in Paris in August 1789 and spent his entire life in a struggle against the legacy of that turbulent summer.[3] His father, Louis-François Cauchy, had been the second in command of the Paris police, but lost his position in the aftermath of the storming of the Bastille. Undeterred, the elder Cauchy was soon serving the revolutionary regime first as chief of the Bureau of Almshouses and, after 1795, as director of the Division of Crafts and Manufacturing in the Ministry of the Interior. His fortunes continued to rise as regimes changed: in January 1801 Louis-François declared himself an enthusiastic Bonapartist and was rewarded by being elected to the combined posts of secretary general, archivist, and keeper of the seal of the newly created Senate. Faithful civil servant that he was, and flexible in his loyalties, the elder Cauchy retained his high position through the Empire (1804–1814), the Bourbon Restoration (1814–1830), and Louis-Philippe's "July Monarchy" (1830–1848). He died within months of finally losing his post following the monarchy's fall in 1848.[4]

As the son of the secretary general, young Augustin-Louis spent much of his time in the Palais du Luxembourg in Paris, which was home to the Senate. There he became acquainted with Pierre-Simon Laplace, who was president of the Senate and his father's direct superior. Laplace was the leading mathematical physicist in Europe, and he took an interest in the young Cauchy, who was already showing a keen mathematical talent. Augustin-Louis also made the acquaintance of the aging senator and mathematician Joseph-Louis Lagrange, the man who more than any other influenced his approach to mathematics. Thanks to his father's position, even as a boy Cauchy was already at home in the company of the leading mathematicians of the age, the ones best positioned to smooth the path of his future career.

Cauchy studied hard for the entrance examinations to the École polytechnique and was admitted with little trouble. He began his studies there in late 1805, and after a difficult first year, in which his performance was mediocre, his natural mathematical talents and his exemplary work ethic reasserted themselves. He graduated third in his class in 1807 and moved on to advanced training in road and bridge engineering. It was during his two years at the École des ponts et chaussées (School of Bridges and Roads) that Cauchy established himself as a rising mathematical star. He won four first-place prizes in school competitions in his first year and two more in his second year, as well as several second- and third-place prizes. He might have done even better had not two of his second-year essays been lost by the examiners, making him ineligible for the prizes. It was an ironic foreshadowing of the fate of Galois' and Abel's memoirs that were entrusted to Cauchy's care years later and goes some way to explain why he did not consider his carelessness a capital offense. Cauchy's lost essays were discovered years later among the papers of one of the examiners, Marie-Riche de Prony. If the troubled relationship between Cauchy and Prony in later years is any indication, one may wonder if the disappearance of Cauchy's papers in 1809 was wholly accidental.

Unlike Galois, however, Cauchy spent little time worrying about dark forces working against him. In 1810 he was appointed to a prestigious position for a young engineer, working on the construction of the new naval harbor in Cherbourg. During the two years he spent in that Atlantic port, he proved himself an excellent engineer both in theory and in practice. He also started exhibiting the social and moral rigidity for which he became known in later years, and that was to cause him no end of trouble. Louis-François Cauchy had taken pains to introduce his son to the leading citizens of Cherbourg, but Augustin-Louis soon found their company tiresome. Worse, he made no effort to hide his dislikes and was unconcerned about giving offense to powerful connections who had welcomed him as a friend.

The reason was that Augustin-Louis, like the rest of the Cauchy family, was a devout Catholic. Unlike his father, however, who successfully adapted himself to four successive regimes of varying religiosity, the young Cauchy had no talent or desire to accommodate the prevailing political winds. Already while he was a student at the École polytechnique, which was dominated by anticlerical republicans, he had joined the Congrégation de la Sainte Vierge, a semisecret organization founded by a Jesuit priest to fight irreligion and work for the return of Catholicism to France. A few years later, in Cherbourg, he showed increasing disdain for the local elite and what he considered their frivolous mores. When his parents wrote to warn him of the aloof reputation he was earning himself in the town, he yielded not an inch: "So they're claiming that my devotion is causing me to become proud, arrogant, and self-infatuated?" he wrote in response. "Exactly who is making these claims? Not people who have much religion themselves."[5]

Cauchy spent two years in Cherbourg, during which he increasingly devoted his time to abstract mathematical work rather than practical engineering. He sent several papers to the Academy of Sciences in Paris, all of which were well received, and one of them—on polygons and polyhedrons—with enthusiasm. When he returned to Paris in 1812, Cauchy continued working on public engineering projects, but his heart was set on a different career, that of a full-time scholar. That meant membership in the Academy of Sciences, or the "first class" of the Institut national des sciences et arts, as it was called under the Empire, and Cauchy made every effort to be admitted into its ranks. Three times between 1813 and 1815 Cauchy put his name forward as a candidate for the Institut. Each time he did not hesitate to make use of his father's influence and connections, as well as his personal acquaintance with some of the leading members of the Institut to bolster his case, all to no avail. In 1813 Louis Poinsot was elected with 23 votes, whereas Cauchy was eliminated in the first round, having received only 2 votes. In 1814, after intense campaigning on his behalf by Laplace and the venerable naturalist Georges Cuvier, Cauchy gathered 10 votes, whereas the winner, André-Marie Ampère (1775–1836), had 23. Finally, in March 1815 a seat opened unexpectedly when Napoleon, having returned from Elba, resigned his position in the Institut. This time the winning candidate, Pierre Molard, received 28 votes, whereas Cauchy had not a single vote cast on his behalf.

Why was Cauchy repeatedly rebuffed in his efforts to gain a position at the Institut? Certainly there was no question of his qualification for membership, for Cauchy was a rising star in mathematical circles, and his talents and accomplishments were well known to the voting members of the Institut. Laplace may have been influenced by his friendship with the elder Cauchy when he wrote before the 1814 elections that "no one is more qualified to fill this important position than that young man," but his endorsement of young Cauchy was well deserved.[6] Yet not only was Cauchy not elected to the Institut, but in three successive elections he did not even come close.

Certainly some of Cauchy's competitors in the elections, such as Poinsot and Ampère, were scholars of the first rank and deserved a position in the Institut as much as he did, but that is not sufficient to explain Cauchy's very poor showing in election after election. One likely factor was political prejudice, since most of the members of the Institut were loyal republicans, whereas Cauchy made no secret of his royalist inclinations and religious piousness. The elections of 1815, in particular, which took place during the ferment of the 100 days of Napoleon's return to power, were an inauspicious time for a man of Cauchy's political leanings to put his name forward. Then there was the not-so-subtle interference of Cauchy's father, the secretary general of the Senate, who unapologetically used every means at his disposal to promote his son's candidacy. It seems likely that Louis-François overplayed his hand in attempting to influence the elections, which led to a backlash against his son by resentful members of the Institut. Finally, there was the fact that the young Augustin-Louis had already gained a reputation for aloofness and arrogance and was not well liked. This could be seen as a serious problem for members of the Institut, who had to cooperate closely in reviewing papers submitted to them and regularly sit on committees together. They would think twice about adding to their ranks a scholar, however brilliant, who was thought to be personally rigid and unaccommodating.

In 1815 Cauchy was nearly the same age as Abel was at the time of his death in 1829. From a material standpoint, Cauchy's life was much easier, for his well-to-do family and his employment as an engineer by the state ensured that he would never suffer the economic hardships that plagued Abel. But from the standpoint of his mathematical reputation, there is no question that at the same age Abel was far ahead of Cauchy. At the age of 25 Abel had published groundbreaking work in the leading mathematical journals of the time. His articles were acknowledged as works of genius by leading mathematicians, and his fame was growing fast not only in his own country but all across Europe. Most significantly, he was being actively recruited for a position in Berlin, a rising center of mathematical research, where well-connected people were using all their influence to create a position just for him. Cauchy's stature at the same point in his life was far more modest. He had published several articles and had presented a few papers to the Institut, and some of his work had gained high praise from its members. But his repeated attempts to establish himself as a professional mathematician had seemingly hit an impenetrable wall of resistance.

In fact, Cauchy's disappointments at the hands of the French scientific establishment at this point in his life would seem to rival Galois' travails a decade and a half later. Both Cauchy and Galois had suffered the injury of having their papers "accidentally misplaced" by senior mathematicians, and both had been repeatedly rebuffed in their approaches to the Institut. Both, furthermore, suffered for their unconventional political convictions, which were nearly mirror images of each other—Galois a radical republican, Cauchy a royalist reactionary. All of which is to say that in 1815 the young Cauchy, far from being the "establishment mathematician" of later lore, had as much cause for resentment against the established mathematicians as the young Galois would have in 1832. As we have seen, Galois channeled his resentment toward self-destruction and early death, after which it formed the basis of his lasting legend as a victim of an uncaring world. Cauchy was more resilient, and year after year, despite repeated setbacks, he put his name forward for membership in the Institut.

We will never know how long Cauchy would have persisted in his attempts in the face of repeated rejection, or whether resentment would ultimately have gotten the better of him, as it did of Galois. Cauchy did not possess Galois' flair for the dramatic or his extreme self-destructive bent, but one might wonder whether a few more years of being turned away from the Institut would have persuaded Cauchy to seek a career in a different field. He could have turned to the study of law, which was the Cauchy family tradition, or, being a devout Catholic and a firm believer in the privileged status of priests, he might have become a clergyman himself. But in 1815, the year in which Cauchy received not a single vote for membership in the Institut, Napoleon's empire finally crumbled, and a Bourbon king once more reigned in France. For Cauchy, almost overnight everything changed.

The Man of the Hour

For the Restoration regime that came to power in 1815, Cauchy appeared as a solution to a pressing dilemma. Louis XVIII and his advisors were duly impressed with the achievements of French mathematicians and natural philosophers during the revolutionary decades. In 1815 Paris was the undisputed center of scientific learning in Europe, and the king and his ministers wanted to keep it that way. Unfortunately, however, the majority of the scholars who deserved credit for French scientific preeminence were highly suspect in the eyes of the new regime: most had republican leanings and had often been active revolutionaries themselves. As much as possible, the new regime sought to replace the most disloyal members of the scientific community without sacrificing the overall quality and standing of French science. Cauchy, who was unquestionably a brilliant mathematician but who was also an outspoken royalist whose career had suffered because of his political leanings, was a godsend to them. Here was a brave and loyal royalist who was worthy of taking the place of the disloyal revolutionary savants.

Thanks to these favorable political winds, the Restoration era from 1815 to 1830 proved to be the high point of Cauchy's career. In this period of 15 years he presented no less than 92 papers to the academy and published 41 articles in scientific journals, as well as 10 longer works published separately. Then, as later in his career, Cauchy was extremely prolific, and his interests ranged over a wide variety of topics in mathematics and mathematical physics. But it was during the years of the Restoration that Cauchy accomplished the work for which he is best known, and that has had a profound effect on the course of mathematics since his day: his work on the foundations of the calculus, including the famous δ-ε (delta-epsilon) formulation, which is taught to students of mathematics to this day. This new formulation, as we shall see, encapsulated a whole new view of the nature of mathematics and its relationship to the world, and Cauchy became the standard-bearer of the new approach. During the same years Cauchy also enjoyed the public accolades that had been denied him previously and were not to be his again in later years. His outspoken and unyielding political views had made him the odd man out in the Napoleonic Empire, but under the new regime he became, quite literally, the darling of the authorities. The young mathematician was soon enjoying the rewards of his new status.

The first of these came only a few months after the king's return to France, when in December 1815 Cauchy was unexpectedly made assistant professor of analysis at the École polytechnique. Cauchy's appointment to a position at the school where he had been a student less than a decade earlier was highly unusual, for he had not even been a tutor *(répétiteur)* there, a position that was usually a prerequisite for becoming a regular teacher. The appointment was also an affront to Poinsot, who was the regular professor of analysis at the school, but who could not dispense his teaching duties because of alleged illness. Poinsot had wanted a substitute to take on his teaching responsibilities, and Cauchy's regular appointment was clearly an encroachment on his position. Despite all this, Cauchy's appointment went through without a hitch, and he began teaching the second year of analysis at the École polytechnique that winter. A few months later the king's cabinet felt secure enough to take on the subversive radicals of the École polytechnique head-on. The École was closed for six months, during which time a committee headed by Laplace reorganized the school along lines more friendly to the new regime, purging several of the more vocal republicans on the faculty. Cauchy was once again the beneficiary of this political interference, for among those dismissed was the republican Poinsot. At the age of 27 Cauchy became full professor of analysis at the École polytechnique.

A full professorship at the most prestigious institution of learning in France secured for Cauchy a solid income, as well as a place among the Parisian scientific elite.[7] What he truly wanted, however, was membership in the Academy of Sciences, as the former first class of the Institut was now called once again. In 1815 Cauchy published two groundbreaking studies, one on the theory of light, the other on Fermat's conjecture on polygonal numbers, both of which established his credentials as a worthy candidate for the academy. Then he bided his time in the hope that the changed political atmosphere would help reverse the repeated rebuffs he had previously suffered.

He did not have long to wait. The Restoration regime was now moving quickly against its enemies, and high on its list was the purging of savants who were considered loyalists of the Revolution and the Empire. Among the most obvious targets were the academicians Lazare Carnot and Gaspard Monge, both leading mathematicians with unquestionable credentials, as well as high officials in the revolutionary and imperial governments. Carnot had been a member, alongside Robespierre, of the Committee of Public Safety and later a member of the Directory, and was credited with the military reforms that made possible the victories of the revolutionary army. Monge had been a minister of the navy under the Revolution, a member of Napoleon's Egyptian expedition, and a personal friend of the emperor. Most significantly, from 1797 to the restoration of Louis XVIII, Monge had been the director of the École polytechnique and had left his personal imprimatur on the teachings and the culture of the school. On March 21, 1816, Louis XVIII announced the expulsion of Carnot and Monge from the academy. In their place were appointed Louis Bréguet and Augustin-Louis Cauchy.

So, in the end, Cauchy never was elected to the academy but was appointed by royal fiat. It is an irony of fate that the mathematician who was perpetually rejected for membership in that august institution and required gross outside interference to be entered into its membership rolls later came to be seen as the embodiment of institutional French science. That is how Abel and Galois saw him, and that is the image that has come down to us through the centuries. But Cauchy's reality was very different: he was an outsider, snubbed and shunned by more established mathematicians. When the opportunity came to gain admission to the academy without submitting to the indignities of a vote by members who seemed prejudiced against him, Cauchy did not hesitate. He accepted the royal appointment without question and promptly took his place in the academy. The position, he thought, was his by right of his mathematical accomplishments, whether the other members of the academy liked it or not.

If Cauchy was not well liked by his scientific peers before his appointment, his acceptance of government interference in the affairs of the academy did nothing to increase his popularity. The royal intervention was an insult to all members of the academy, even to those few who were in other respects friendly to the new regime. Carnot and Monge, furthermore, were highly respected and well-connected members of the academy, and their cavalier dismissal was a slap in the face of the institution. Cauchy could not have cared less, and he made no apologies for entering the academy under such circumstances. But he paid a heavy price in the esteem of his peers: "Cauchy," wrote academician Joseph Bertrand, "found few defenders. He has seen more than one friend who, though naturally tolerant and decent, turned away and refused to call him 'brother.' "[8] For the time being Cauchy could shrug off the hostility of other academicians, secure in the knowledge that his political standing made him invulnerable to their attacks. But there would come a time when his failure to cultivate good relations with his colleagues would return to haunt him.

In the meantime, honors and positions continued to accumulate for Cauchy, who was, politically at least, the man of the hour. In 1819, at the age of 30, he was awarded the Légion d'honneur by Louis XVIII, and in the following years he added two prestigious teaching positions to his professorship at the École polytechnique. Since 1817 Cauchy had been doing replacement teaching in the Collège de France for the professor of mathematical physics, Jean-Baptiste Biot (1774–1862). In 1824, when Biot took a prolonged leave of absence, Cauchy became his full-time substitute, a position he retained until the revolution of 1830. At about the same time, in 1821, Cauchy was appointed as Poisson's replacement in teaching mechanics at the Faculty of Sciences at the Sorbonne (the University of Paris). Thus in the 1820s Cauchy simultaneously held three full-time teaching positions at some of the most prestigious and demanding institutions of learning in Europe. Add to that the fact that as a member of the academy, he was required to sit on dozens of committees charged with reporting on works presented to the institution, and that on top of that he was regularly publishing articles and books about his own research, and one gets a taste of the almost superhuman dedication and discipline that Cauchy brought to his career. This also goes some way to explain why Poncelet, Abel, and Galois, who encountered Cauchy during these years, all felt that he seemed too busy to spare any time for them: quite obviously, he really was.

The Renegade Professor

At the École polytechnique, meanwhile, the new professor of analysis had hit turbulent waters. Given the manner of Cauchy's appointment to the École and to the academy, this was only to be expected, for there was no denying that Cauchy had been imposed on both institutions. At the École the ultraconservative Cauchy had been given preference over more liberal candidates despite never having been a tutor at the school; even more egregiously, at the academy Cauchy was appointed to replace the ousted Monge, who had been the longtime director of the École polytechnique and was still admired, perhaps even revered, by both the faculty and the students.[9] The fact that Cauchy promptly accepted both positions as his rightful due, paying no heed to the prevailing views within the institutions, surely did not endear him to his new colleagues. But there was more: Cauchy was not only politically at odds with most of his colleagues but also sharply disagreed with them over the proper curriculum of his courses at the school.

The dispute between Cauchy and his colleagues about the proper curriculum at the École polytechnique reflected a profound philosophical rift over the nature of mathematical learning that was gradually manifesting itself in this period. When the École was founded during the Revolution, its founders brought with them an understanding of the nature of mathematics they had inherited from the Enlightenment. Mathematics, as understood by the grands géomètres of the eighteenth century and their heirs, was inseparable from the physical world around them. This meant that in studying mathematics they were, in fact, studying the physical world and its hidden structure. Conversely, if a mathematical theory correctly described physical reality, then the theory was essentially true and required no further demonstration. The École polytechnique, an engineering school where advanced mathematics was studied for the express purpose of using it in the natural world, was the institutional embodiment of this Enlightenment view.[10]

But the early nineteenth century saw the emergence of a new and different approach. In this view mathematics was its own insular world, answerable only to its own internal standards of logical coherence and rigor. Accordingly, mathematical statements were completely independent of any external reality; their truth value depended only on correct derivation from simpler mathematical truths, not on whether they corresponded to physical reality in the outside world. Certainly mathematics could and should be profitably applied in scientific investigations of the natural world. But before this could be done, mathematics had to be first and foremost true to itself, or it would risk falling into serious error. Cauchy was one of the earliest adherents to this view of mathematics, as were Abel and Galois in the following years. In fact, more than any other mathematician of the period, Cauchy is identified with this new approach.[11]

Cauchy moved quickly to try to shape the curriculum at the École polytechnique according to his views. Already in 1816, after only one year of teaching, he proposed a fundamental reform to the school's Curriculum Committee (Conseil d'instruction). Cauchy was dissatisfied with the fact that mathematics was taught alongside mechanics, mainly as a tool to solve practical problems in mechanics and engineering. He proposed instead to devote the entire first year of study to an in-depth course in mathematics, leaving mechanics and applications to the second year. This was perfectly in line with his views on the nature of mathematics and its role in the physical sciences and engineering, but it went completely against the traditions of the École, which viewed mathematics as a handmaiden of the sciences, whose sole purpose was to aid in the understanding and management of the physical world. Unsurprisingly, Cauchy's proposals were rejected by the committee.[12]

Faced with an institutional rebuff, Cauchy attempted to implement his reform on his own. Because he could not control the teachings of the other professors, he simply ignored them and proceeded to shape his own courses in accordance with his views. As a result, Cauchy's 15 years of teaching at the École polytechnique turned into a prolonged game of cat and mouse in which the school ever more urgently and desperately tried to control the teachings of the renegade professor of analysis. Cauchy, for his part, stalled, argued, and occasionally went through the motions of accommodating his critics, but he never gave up his ultimate plan of teaching pure mathematics on its own terms, separate from its applications. Cauchy could do this because no matter how frustrated the school's administration grew with his obstinate refusal to follow the curriculum, it could not get rid of him. As long as a Bourbon king reigned in France, the conservative royalist Cauchy was untouchable.

Early signs of trouble first appeared in 1819 when François Arago (1785–1853), the professor of applied analysis, and Alexis Petit (1791–1820), professor of physics, complained of the inadequacy of the students' training in the calculus. Cauchy, they claimed, was spending too much time teaching abstract mathematics that the students would never have occasion to use and too little time equipping them with useful mathematical tools for the engineering careers that lay before them. The director of the École added his voice to the complaint, charging that Cauchy's teaching of pure mathematics was "an uncalled-for extravagance" that was "prejudicial to the other branches." He then demanded that Cauchy adhere closely to the official syllabus.[13]

The director's orders seemingly had little effect, for the following year the school's Conseil d'instruction felt compelled to demand that Cauchy and his friend and ally Ampère revise their courses. In response, Cauchy assured the committee members that all their concerns would be laid to rest with the forthcoming publication of the first volume of his *Cours d'analyse de l'École royale polytechnique* (The Course in Analysis of the École Polytechnique).[14] When this work, commonly known as *Cours d'analyse,* was published in 1821, it was indeed a mathematical masterpiece. It is Cauchy's most famous work and ultimately changed the course mathematics was to take in the following two centuries. But it is hard to believe that Cauchy really thought that it would assuage his critics. It was, in fact, a straightforward exposition of the course in analysis as Cauchy would ideally have taught it—beginning with simple assumptions and definitions and proceeding through rigorous deduction to ever more advanced results. It made no concessions to the practical engineering approaches favored by the École's director and most of the other professors. Far from reassuring Cauchy's colleagues, it confirmed their worst fears that Cauchy was intent on defying the school's official curriculum and continuing to teach his pure brand of mathematics. They did their best to ensure that *Cours d'analyse,* the most famous "course" in the history of mathematics, was never actually taught.[15]

Many of the students were no more satisfied with Cauchy's course than were the school's faculty.[16] In April 1821, a few months before the publication of *Cours d'analyse,* Cauchy was booed by several students after his class had gone overtime. This was a highly unusual occurrence at the École polytechnique, and the school's director, Baron François-Louis Bouchu (1771–1839), submitted a report about it to the minister of the interior. Bouchu, a reserve general, blamed the students for their "thoughtless insult" and demanded that the class leaders apologize to Cauchy. More surprisingly, however, Bouchu assigned equal blame for the incident to Cauchy himself, arguing that his refusal to teach the syllabus and his habit of routinely extending his lectures beyond their allotted time were causing increasing resentment among the students. "I can no longer hide the fact that for the past five years he has been given many warnings to simplify his teaching methods so as to bring them into line with the official program," Bouchu wrote.[17] "Though he [Cauchy] is not the only professor here who possesses superior abilities and excellent principles, he is the only one who perseveres in disregarding the official syllabus," he added in a later report.[18]

The director's strong words did not move the minister of the interior, who viewed the affair as essentially political. Cauchy, in his view, was being harassed by both faculty and students not because of his teaching methods, but because of his political views. There was reason for the minister's suspicions, for the royalist Cauchy was hardly a popular figure at a school famous for its strong republican leanings. Furthermore, Cauchy was always known at the École as the man who had taken the place of the much-beloved Monge after his ouster from the academy. The minister was therefore determined not to permit any sanctions against Cauchy or to give his support to any attempt to undermine the position of the loyal professor of analysis. As long as the current regime remained in power, Cauchy, it seems, could not be disciplined.

Unable to rid themselves of the defiant professor of analysis, the members of the Conseil d'instruction of the École did their best to curb his independence. When Cauchy returned to teaching his class in 1821, he was under orders to submit written summaries of his lectures for review. Cauchy did so, and in 1823 he even published his lectures. But in content this new work, known as *Le calcul infinitésimal,* was basically a continuation of the *Cours d'analyse* of 1821.[19] As before, Cauchy continued to develop his groundbreaking mathematical views on the foundations of the calculus; and as before, his colleagues at the École continued to insist that such abstract theorizing had no place in the training of future engineers.

The conflict reached such a pitch that in 1823 the minister of the interior felt compelled to appoint a governmental commission to investigate the goings-on in the classrooms of Cauchy and his friend and fellow conservative, Ampère. The commission was chaired by Cauchy's former patron, Laplace, and also included Poisson and Prony, the man who years earlier had lost Cauchy's essays among his papers. In this case, however, all three were in concert in condemning Cauchy's conduct. They demanded that the two wayward instructors submit their lesson plans for the first semester of 1824 to the commission, and whereas Ampère complied, Cauchy stalled for time. He first argued that he had no time to produce the work required by the commission and later claimed that printing difficulties were preventing him from complying with the commission's requests. When Cauchy finally did submit a partial report on his teachings, Laplace was appalled, commenting that "some of the material handed in by one of the professors of analysis was so unintelligible to him that he could only make sense of it after a third reading."[20]

Despite his friendship with Cauchy's father, Laplace made it clear that his support for Augustin-Louis was at an end. This was understandable, for whatever his political loyalties, Laplace worked very much in the Enlightenment tradition that viewed mathematics as inseparable from its manifestations in the natural world. For him, Cauchy's abstractions were not only bad pedagogy but also meaningless sophistries that drew mathematics away from its proper subject matter—physical reality. Unfortunately for Laplace and the École's administration, there was little that the commission could actually do to enforce its judgments on Cauchy, and the chorus of complaints against him continued. In 1825 Arago, the professor of applied analysis, and Jacques-Philippe Binet (1786–1856), the École's inspecteur des études, both complained about the level of practical mathematical training attained by Cauchy's students. This led to a complete overhaul of the analysis curriculum at the school that flatly rejected Cauchy's approach to teaching the calculus.

The curriculum reform was aimed specifically against Cauchy's practice of teaching a sophisticated and rigorous approach to the calculus based on the concept of the limit. This method, known since Cauchy as the δ-ε (delta-epsilon) approach, is still in use today and is taught in college-level mathematics courses.[21] The new curriculum at the École, however, banned it and replaced it by the more traditional "infinitesimal" approach to the calculus. Infinitesimals are notoriously problematic from a conceptual point of view, for they entail the paradoxical notion that the continuum is composed of an infinite number of segments of zero magnitude. But it appeals to material intuition in a way that Cauchy's rigorous approach cannot, and it was therefore used almost exclusively by practicing engineers. Furthermore, it strongly suggested the mutual interdependence of mathematics and the physical world, a view inherited from the Enlightenment and shared by the majority of the faculty at the École.

Cauchy was certainly not opposed to applying mathematics to physical problems, and he dedicated much of his scientific work to that goal. He responded to the curriculum committee's rebuke by publishing a book on the application of analysis to geometry, *Leçons sur les applications du calcul infinitesimal à la géométrie* (1826).[22] But when it came to the teaching of analysis itself, Cauchy was adamant and continued following his own curriculum. Cauchy's critics at the school proved equally determined to curb the independence of the defiant professor, assigning a colleague to oversee Cauchy's teachings. For the task they chose none other than Cauchy's old nemesis, Prony, who for the remainder of Cauchy's tenure at the École submitted highly critical annual reports on his recalcitrant charge.[23]

In 1826 the increasingly frustrated school authorities convened yet another committee to deal with the goings-on in Cauchy's analysis classes. Laplace, who once again chaired the committee, complained that as president of the commission appointed by the minister of the interior, he had tried for three years to obtain from Cauchy satisfactory written material on his courses, but "it has been impossible . . . to get him to comply with the Conseil's instructions and the Minister's decision."[24] Following up on Laplace's strong words, the École's governor for the first time threatened tough disciplinary action: "In case other attempts are made, and are unsuccessful, then he would consider himself as obliged to propose to the government that it take harsh measures, which he would find very distasteful."[25]

Even these strong words, however, proved to be empty threats. The governor, as his statement makes clear, knew that he could not act without support from the government. This was clearly lacking, and Cauchy continued his insubordination unmolested. The school's faculty, in the meantime, was growing increasingly frantic at having in its midst a rebellious but untouchable professor. Tired of waiting for Cauchy to deliver his lesson plans for inspection, it decided in February 1828 to take the extreme measure of placing stenographers in Cauchy's classroom. The idea was that their notes would be the basis of the text of Cauchy's courses, which would then be published. But this plan to bring the professor of analysis to account failed, like all its predecessors.[26] Cauchy sent a letter of protest to the minister of the interior, claiming that he did not have time to correct the stenographers' texts, and the school backed away from the plan. Evidently Cauchy could still count on support and goodwill in high places.

Meanwhile, the school administration's efforts to make things as uncomfortable as possible for Cauchy produced an unexpected effect: Ampère, who was to be monitored along with Cauchy, had had enough and resigned his position at the École in May 1828. Cauchy, however, proved immune to the hostility of his colleagues and retained his position right up to the revolution of 1830. Only then, deprived of the support of his friends in the government and no longer untouchable, did Cauchy finally give up his position at the École.

The Martyr

In the same years in which Cauchy was turning into a pariah at the École polytechnique, he was also increasingly raising the ire of his peers at the academy. Cauchy, a true believer in the Catholic and royalist causes, seemed to be using his eminent position to promote his views publicly at every turn. In July 1824 he went far beyond the boundaries of his specialty in responding to a report by anatomist Étienne Geoffroy Saint-Hilaire (1772–1844), who was a supporter of Lamarck's evolutionary theories and believed in the transmutation of species. To conservatives such as Cauchy, these views led directly to materialism and atheism, so few were surprised that the devout Catholic mathematician rose to denounce Geoffroy's report. Cauchy, however, did not stop there but used the opportunity to condemn the theories of Franz Joseph Gall (1758–1828), the founder of phrenology, accusing him of promoting "a principle that rejects both the true philosophy and the vital doctrines on which rest the peace and well being of society."[27] Although most academicians were not supporters of Gall, whom they never accepted for membership, they were understandably irritated at Cauchy's blatant use of the academy to promote his own principles. But Cauchy was not done. In October of the same year he responded to a report on the theory of light that mentioned in passing that Newton had doubted the existence of the soul. Not so, thundered Cauchy. The error, he showed, could be traced to Voltaire, and he went on to attack that Enlightenment icon and reaffirm the superiority of the Christian religion.[28]

Cauchy's clericalist pronouncements earned him not only the hostility of his fellow academicians but also public ridicule in the republican press. One article that followed his attack on Voltaire asserted that he was delivering religious sermons rather than scientific presentations in the academy. No less an eminence than the author Stendhal claimed that Cauchy was "a veritable Jesuit in short frock" and that his religious pronouncements were greeted with bursts of laughter by the academicians. "This courageous man," he wrote mockingly, "apparently wants to be a martyr to contempt."[29]

There was more truth than Stendhal intended in his characterization of Cauchy, for in his own eyes Cauchy was indeed a martyr. His rigid and unbending adherence to unpopular doctrines and his persistence in publicly advocating them in the face of mounting opposition earned him no end of trouble. But Cauchy was willing to pay that price for his steadfast devotion to the truth as he saw it—political, religious, and scientific. What comes across from Cauchy's tumultuous career at the École polytechnique and the academy is hardly the image of an establishment savant, comfortable in his secure position and haughtily pronouncing on the future of aspiring mathematicians. It is, rather, the picture of a man under siege, incessantly harassed for his political convictions, as well as his mathematical views. To outsiders such as Abel and Galois, the academician Cauchy, with his teaching appointments at all the greatest centers of learning in the country, naturally appeared as an unassailable embodiment of institutional power. But that was not the reality of Cauchy's life as he experienced it. Despite his universally acknowledged mathematical brilliance, Cauchy was personally, politically, and professionally isolated. He was, in his own eyes, a martyr.

Cauchy's true martyrdom began in July 1830, when three bloody days of insurrection spelled the end of Bourbon rule in France. The regime that had steadfastly supported Cauchy came crashing down, carrying with it his unassailable position in the French scientific establishment. While the frustrated Galois raged within the locked gates of the École normale, his peers at the École polytechnique broke out and joined the uprising. To Cauchy's horror, his own students took to the barricades, led bands of insurgents throughout the city, and played a vital role in the outcome of the insurrection. By the end of the month King Charles X had left Paris, and a week later Louis-Philippe, Duke of Orléans, was proclaimed king of the French. Although many radicals bemoaned the substitution of one king for another as a betrayal of the republican cause, the new regime was in fact profoundly different from its predecessor. Whereas the Restoration regime had been legitimist and clericalist, with strong absolutist leanings, Louis-Philippe's "July Monarchy" was a limited constitutional monarchy and fundamentally secular. To Cauchy, it represented a resurgence of the hated ideals of the Revolution and the destruction of the world order in which he believed. When a law was passed in late August requiring that all civil servants take an oath of loyalty to the new government, Cauchy hurriedly left Paris before he would be required to swear his allegiance.

Exile

Whatever Cauchy's intentions were when he left Paris, in practice his leave of absence turned into eight years of self-imposed exile from France.[30] Cauchy was a family man now, having married Aloise de Bure in 1818, and the father of two young daughters, but in early September he left his family in Paris and headed to the town of Fribourg in Switzerland. It would be four years before Cauchy's family joined him in his exile.

It did not take nearly this long for Cauchy to lose his standing in the French scientific community. Within a few months of his departure, nearly all his distinguished positions in the French capital evaporated. In November he lost his professorship at the Faculty of Sciences of the University of Paris, and in March 1831 he was stripped of his rank of engineer of bridges and roads *(ponts et chaussées)*. The cruelest blow undoubtedly came in February 1831, when he was summarily dismissed from his professorship at the École polytechnique. One can only imagine the relief of the faculty and administration at the École when they were finally able to dispose of their insubordinate colleague, but for Cauchy, this victory of his critics was a bitter pill to swallow. Ironically, Cauchy did retain the one position to which he had been personally appointed by Louis XVIII—his membership in the Academy of Sciences. Because this was not a civil service post, and because the academy had many foreign members, no oath of loyalty was required of its members, and there were no grounds for dismissing Cauchy. Furthermore, although Cauchy had never been elected to the academy, no one doubted that he was eminently qualified for the position, and he was left in peace.

There is no question that being deprived of his positions was hard on Cauchy. Although he never bothered to ingratiate himself with his colleagues and appeared indifferent to what they thought of him personally, he nevertheless attached great value to formal titles and positions. This was evident already in the manner in which Cauchy had obtained his membership in the academy in 1816. Other scholars might have felt that a political appointment to the academy, which did not reflect the esteem of their colleagues, was an empty shell. But not so Cauchy: his prompt and unapologetic acceptance of the position indicates that he valued the prestige of being an academician for its own sake, regardless of the opinions of his fellows. The same is true of Cauchy's professorship at the École polytechnique, which was forced on the school, but which Cauchy accepted without reservation and treated as his rightly due. In fact, Cauchy's very insistence on collecting a plethora of honors and prestigious affiliations throughout this period, even to the point of risking his health by overwork, points to his inordinate love of such distinctions. Their sudden loss in the months after his exile was unquestionably painful to him.

And yet Cauchy persevered. If he had remained in Paris and sworn the required oath, he might well have retained at least some of his positions. He could have followed the example of his father, who shared his son's political outlook but nevertheless swore allegiance to Louis-Philippe and retained his post at the Chamber of Peers (as the Senate was now called) for another 18 years.[31] But the younger Cauchy never considered this. The only legitimate French regime, for Cauchy, was an absolutist monarchy supported by the Catholic Church, with a Bourbon king at its head. Louis-Philippe was an illegitimate monarch who was also a puppet in the hands of revolutionary republicans. There could be no compromise with such a regime, and swearing an oath of loyalty to it was out of the question. It was a matter of standing up for the truth, and Cauchy was willing to suffer for such matters. A few years earlier Stendhal had derided him as "a martyr to contempt" for his knack for bringing abuse upon himself. In his own eyes Cauchy was indeed a martyr, but to truth—the truth of the divine right of kings, the truth of the Catholic religion, and the truth of mathematics.

In the summer of 1831, after the political climate in Fribourg turned hostile, Cauchy moved to Turin, where he joined a legitimist community living under the protection of King Carlo Alberto of Sardinia. Cauchy stayed there for two years, during which he collaborated closely with scholars at the Turin Academy and also taught at the local university, which offered him a chair. But in the summer of 1833, as he was weighing his options, Cauchy received what for him was an offer he could not refuse. The tutor of the Duke of Bordeaux, grandson of the exiled King Charles X and legitimist heir to the throne of France, wrote to Cauchy and asked him to be the young prince's tutor in the sciences. Ever ready to aid the Bourbon cause, Cauchy packed his bags and headed to Prague, where the court in exile had made its home.

If Cauchy ever earned the label "martyr," it was during the years he spent as tutor to the Duke of Bordeaux. The young prince, who was thirteen when Cauchy arrived in Prague, showed no scientific or mathematical aptitude. Cauchy, who in his days at the École polytechnique had often been accused of teaching only to the top students, now had in his charge a student who had difficulty mastering even the most elementary concepts. Dutifully, Cauchy set out to compose treatises on elementary arithmetic for the sole use of the young duke, and with inexhaustible patience he explained the same simple principles over and over again, all to no avail; the prince had no liking for the mathematical sciences and even less for his devoted teacher. He particularly enjoyed playing practical jokes and humiliating Cauchy, who always responded with a humble inquiry as to how he had displeased his royal student.[32]

But through it all Cauchy persevered. He managed to retain his position for five long years in a court rife with intrigue, in which many other officials, functionaries, and tutors were regularly dismissed and replaced. Perhaps it was the prestige that having the leading mathematician in Europe in residence brought to the court in exile that made him valuable to the king above and beyond his services to the young duke. Or perhaps it was the straightforward manner in which Cauchy handled himself that helped him survive the turbulent waters of life at court. Far from showing a talent for underhanded dealings in the manner of traditional courtiers, Cauchy never made a secret of his views. He doggedly and predictably supported the most extreme conservative positions and people, and that proved to be a successful formula for longevity in the Bourbon court in exile. From 1833 to 1838, first in Prague and then in Göritz, Prussia, Cauchy endured the abuse of the Duke of Bordeaux and the hostility of rival members of the court to remain the prince's tutor in the sciences. When the old king made Cauchy a baron as a reward for his faithful services to the royal family, there is no question that the long-suffering Cauchy had earned the honor.

Pariah in Paris

In 1838 the Duke of Bordeaux reached majority, ending Cauchy's role at the exiled court. Eight long years of exile had taken their toll on the mathematician, and once he felt that his duty to the monarch had been fully discharged, he decided to return home to Paris. Cauchy had always been close to his birth family, which had remained in Paris, and the desire to be near his parents as they aged likely played a role in his decision. His return certainly did not mean that Cauchy had made his peace with Louis Philippe's July Monarchy. To the contrary, his opposition to the regime was as unbending as it ever was, he remained adamant in his refusal to swear allegiance to it, and he never stopped working toward its overthrow and the return of the Bourbons. But such a restoration appeared a very remote possibility in 1838, and there seemed little point in trying to wait out the exile. Perhaps Cauchy also believed that he could work more effectively for the Bourbon cause from Paris than from abroad. Be that as it may, on October 22, 1838, after eight years of absence, Cauchy once again attended a meeting of the Paris Academy of Sciences. It was the first step in his campaign to recover his position in French scientific life that he had lost in 1830. It would last for the rest of his life.

Early on, Cauchy's prospects seemed promising. Eight years of separation seemed to have had a soothing effect on the attitudes of France's leading mathematicians toward Cauchy, and many of them were willing to welcome their brilliant but difficult colleague back into their ranks. In 1839 Cauchy's old nemesis Prony died, creating a vacancy at the Bureau des longitudes, and Cauchy was immediately considered the leading candidate to replace him. Although its name implies a specific technical competency, the bureau was in fact a highly prestigious institution whose members were nearly always academicians. The position was highly desirable because unlike membership in the Academy of Sciences, an appointment to the bureau also carried with it a respectable salary. In November, with support from his old friend Biot, but also from Arago, his old critic at the École polytechnique, Cauchy was elected to a position in the Bureau des longitudes.[33]

At this point, however, Cauchy's candidacy encountered what turned out to be an insurmountable roadblock. The bureau was technically a part of the French civil service and therefore required that its members swear an oath of loyalty to the regime. But Cauchy in 1839 was just as determined not to swear allegiance as he had been nine years earlier. Flatly refusing to take the oath, Cauchy, although elected to the bureau, could not be confirmed in his position, could not take part in the institution's meetings, and could not be paid. Over the next few years successive ministers of public instruction in Louis-Philippe's cabinet tried to reach a compromise and reduce to a minimum the conditions Cauchy would have to fulfill in order to comply with the loyalty-oath requirement. Cauchy, however, would have none of it. Any arrangement that would dilute his absolute rejection of the regime was to him unacceptable.

For four years Cauchy remained in limbo, an elected member of the bureau who nevertheless could take no part in its activities. In 1843, faced with Cauchy's obstinate refusal to compromise, the members of the bureau had had enough. They formally excluded Cauchy from the bureau and elected another member in his place. For Arago and other members of the French mathematical community, the affair was undoubtedly a sharp reminder of how difficult a colleague Cauchy had been years before in the École polytechnique. In the future they would not be so welcoming.

Cauchy, however, was undeterred. In 1843 the death of the mathematician Sylvestre-François Lacroix (1765–1843) created a vacancy at the Collège de France, and Cauchy submitted his candidacy. Cauchy's rivals were Liouville, who declared that he "would be first to applaud the choice" if Cauchy was elected, and Libri, who was in the midst of his campaign to denude the provincial libraries of France of their treasures. While simulating friendship toward Cauchy, Libri at the same time lobbied feverishly against him with the college's professors, arguing that Cauchy's appointment would be a victory for the hated Jesuits. Whatever the professors may have thought of Libri's character or motives, his campaign proved effective: in the third and final round of balloting Libri received 13 votes, Liouville 10, and Cauchy but a single vote.[34]

The results of the elections caused an uproar in Parisian academic circles. Not only was Libri already widely suspected of stealing France's national treasures, but his stature as a mathematician could in no way compare with that of his rivals, Cauchy and Liouville. His elevation to a professorship at the college was an outrage that could only be attributed to his skill at manipulating the political sympathies and fears of the electors. But however underhanded Libri's methods were, their success hinged on the fact that his accusations against Cauchy were well founded. Ever since his return from exile, Cauchy had been working closely with the Jesuits in an attempt to break the monopoly of secular public education in France. Cauchy's appointment to the Collège de France would indeed have been hailed by the Jesuits as a victory for their cause and would have enhanced the legitimacy of the unsanctioned Catholic educational system they were establishing alongside the official schools. The professors at the college were anxious to avoid such an outcome and voted for the candidate who presented himself as the champion of state-sponsored education. By playing on Cauchy's personal unpopularity and the fears of a Jesuit takeover, Libri was able to accomplish a stunning electoral upset.

The pattern of rejection that Cauchy encountered in those years is very reminiscent of the years he spent as a young mathematician, knocking on the doors of the Institut and being turned away every time. In 1843, as in 1815, Cauchy's reputation for personal rigidity worked against him, as did his unpopular political stances. In both cases, furthermore, he was completely unwilling to compromise or cultivate useful connections. As he had when he snubbed the local aristocracy in Cherbourg in favor of more devout company, or when he refused to teach the assigned curriculum at the École polytechnique, Cauchy in the 1840s saw himself as the champion of truth in a hostile world. Whether political, religious, or mathematical, for Cauchy the truth was always one and could know no compromise. If standing up for it meant that he would be deprived of the esteem and accolades that were his by right, so be it. Although he felt them deeply, Cauchy was willing to suffer the slings and arrows directed at him for championing what was right. This is the way of the martyr, and in his own eyes Cauchy was a martyr for the truth.

As in the past, political upheaval once again came to Cauchy's aid and rescued him from his enforced isolation. The revolution of 1848 put an end to the July Monarchy and did away with the loyalty-oath requirement for civil servants. Immediately Cauchy presented himself at a meeting of the Bureau des Longitudes, claiming that the obstacle to his taking his position there had been removed. The stunned members, who had long since elected his replacement, showed him the door. Cauchy did have more success at the Faculty of Sciences of the University of Paris, where he was reinstated in 1849 as the professor of mathematical astronomy. Gratifying though this success was, it turned out to be the only positive result Cauchy could ever show for his decades-long campaign to regain his standing in the French scientific community, which he had forfeited in 1830.[35]

In addition to sweeping away Louis-Philippe's regime, the revolution of 1848 also swept Libri out of Paris. With his friends in high places gone, a warrant was issued for Libri's arrest on suspicion of looting the libraries he was charged with inspecting. This left the mathematics chair at the Collège de France vacant, and in 1850, just as in 1843, Cauchy and Liouville presented their candidacies. Even without Libri's scheming, things did not go smoothly. Although Cauchy received more votes than Liouville in the first ballot, he still won less than half the total because two members abstained. In the second ballot Liouville did win over half the votes and was duly appointed professor at the college. Cauchy and his allies protested the decision to the minister of public instruction, claiming that Cauchy had been lawfully elected in the first ballot. Unwilling to decide the issue himself, the minister ordered the college to hold a repeat election, but Cauchy, unbending as ever, refused to cooperate. In his view he had already won the vote in the first round of balloting. When the elections were held anyway, he was soundly defeated, and Liouville was confirmed in his post. Cauchy's attempts to force himself on the college through ministerial fiat echoed the manner in which he had won his positions in the academy and the École polytechnique 35 years earlier. In this case, however, it did not help his cause, and it certainly did not add to his popularity with his colleagues.[36]

Cauchy lived for another seven years after the fiasco of the elections to the Collège de France. During this time he continued teaching in the Faculty of Sciences, which turned out to be a more suitable setting for his pedagogical talents than the École polytechnique. His classes were tiny, sometimes with as few as three students, but these were a highly select group, including the most promising young mathematicians of the next generation. Here, instead of struggling with the constraints of an official curriculum, he was free to develop the advanced topics that interested him, in particular his work on the theory of functions and complex analysis. Ever faithful to his religious and cultural values, he remained active in ultra circles, joining with his Jesuit friends in their fight to legitimize Catholic education in France. He also became deeply involved in a project to establish Catholic schools within the Ottoman Empire, in order to bring the Christians of the East under the sway of Latin Catholicism. The schools, known as écoles d'Orient, exist to this day and are the longest lasting of Cauchy's various religious initiatives. In May 1857 Cauchy fell ill with "great rheumatism" and left Paris for his country house in Sceaux, south of the city. For a while the aging mathematician seemed to be recovering, but on May 22 his condition took a turn for the worse. He died in the early morning hours of May 24, 1857, at the age of 67.

Such was the life of the man who, to outsiders such as Abel and Galois, represented the face of the French scientific establishment. Given his lifelong membership in the academy and his teaching positions in the most prestigious institutions in Paris, their mistake is understandable, but the reality was nonetheless very different. Cauchy was never elected to the academy, but was imposed on it after repeated rejections by its members. That was even more the case at the École polytechnique, where Cauchy encountered the active hostility of his colleagues, who viewed him as a representative of the despised Restoration regime. Cauchy himself only increased his isolation by his endorsement of unpopular views on mathematics and its teaching and his stubborn refusal to cooperate with the school's faculty and administration. Even at the height of his career Cauchy was something of a pariah among his colleagues, and when the political upheaval of 1830 gave them a chance, they were happy to rid themselves of his unpopular presence. Even after he returned from exile in 1838, and even after the political obstacles to his career were lifted in 1848, the man widely regarded as the greatest mathematician in Europe never regained the institutional standing that was his due. From his early days as a young mathematician knocking on the doors of the academy to his disillusioned last years when, tired of repeated rejections, he gave up his efforts to regain the ground he had lost, Cauchy was always the odd man out of French science.

Far from trying to accommodate his critics or tone down his unpopular pronouncements, Cauchy became ever more outspoken and intransigent over the years. On issues of religion and politics, just as on questions of mathematics, there could be no compromise; much like Galois at the opposite end of the political spectrum, Cauchy believed that he was standing up for the truth, and compromise was unthinkable. Galois ultimately paid for his intransigence with his life; Cauchy, who did not carry it to quite the same extremes as his younger colleague, paid for it with a loss of status and income. For both, it was the inevitable result of standing up for the truth. In their own eyes they were martyrs.

The Quest for Truth

In 1842 and 1843 Cauchy gave a series of lectures at the Institut catholique (Catholic Institute), one of the many educational and charitable organizations he helped found in his long career of Catholic activism. It was a difficult time for Cauchy, when his efforts to regain his lost positions seemed to be meeting opposition at every turn. But the Institut catholique offered him a friendly and admiring audience, and Cauchy used the setting to expound on his deeply held views about mathematics, science, religion, and society. His lectures there are one of the few places in Cauchy's voluminous writings where he openly discusses these issues, and they provide a window into the sources of his unshakable convictions.

The goal of all science, Cauchy declared in a lecture titled "Recherche de la vérité" (Quest for the Truth), was the pursuit of truth.[37] That, argued Cauchy, was why some scholars measured the skies and the depths of the oceans, why others climbed to the highest peaks or descended to the depths of the earth, and why others investigated the composition of the elements. "If I asked those who attempt these things what is the goal of their arduous researches," said Cauchy, "they would doubtlessly respond to me that it is the conquest of truth."[38] This, it should be noted, is a highly questionable proposition, because scientists engaged in such endeavors are at least as likely to cite the practical benefits of their work as their key motivation rather than the pursuit of some abstract "truth." But for Cauchy, it seemed self-evident that the purpose of all scientific work was to search for truth.

Mathematics and its applications, the fields that most concerned Cauchy, were no exception: "Is it not," he asked, "in the end, to conquer the truth that [one] interrogates algebra, espouses all the resources of analysis, and arranges in a formula in order to grasp the laws that rule the course of the stars, or the vibrations of the last particles of matter?" His answer followed immediately, and it applied to all the sciences: "Yes, without doubt, the search for truth must be the unique goal of all of science."[39]

Cauchy waxed poetic on the divine qualities of truth, whose attainment made any sacrifice worthwhile:

> The truth is an inestimable treasure, whose acquisition is not followed by any remorse and does not trouble the peace of the soul. The contemplation of these celestial charms, of this divine beauty, suffices to compensate us for the sacrifices that we make to discover it; and heavenly bliss itself is nothing but the plain and entire possession of immortal truth.[40]

Not only is truth valuable beyond measure in its own right, but it is also the glue that holds society together and makes civilization possible. "Yes, without a doubt, the reign of truth must be the object of all our prayers, the end to which we direct all our efforts, all our desires, all our meditations. The truth is that salutary ark that can gather the debris of a society ready to disappear into the abyss."[41] "To recap," he concluded, "there is no real progress, no durable happiness, if one does not love and seek the truth."[42]

For Cauchy, it should be noted, the truth was not a vague abstraction, open to different interpretations. Either something was true or it was false, and the goal of science was to know exactly which was which. That, according to Cauchy, was why exactitude was so important. "Is not exactitude an essential, necessary characteristic of all true science?" Cauchy asked in another of his lectures to the Institut catholique, titled "Sur quelques préjugés contre les physiciens et les géomètres" (On Certain Prejudices against Physicists and Geometers).[43] "Is not the object of all science the quest for or the exact knowledge of the truth?" he continued.[44] Indeed it is. Truth cannot exist without exactitude, for it would be indistinguishable from error: "Is not the exact truth that which distinguishes the true from the false philosophy, that which distinguishes history from romance?" Exactitude is therefore inseparable from truth, and like truth it is of divine origin: "Exactitude and order," Cauchy argued, "are one of the characteristics of the Almighty, who disposed of all things . . . by order, by weight, and by measure."[45]

Cauchy's absolutist views on the nature of truth and precision go a long way to explain the difficult path he chose to pursue in his career. They shed light on his unpopular stance on the nature of mathematics and the way it should be pursued, as well as on his seeming intransigence and his unwillingness to concede an inch even when a bit of compromise was clearly in his own best interests and those of his cause. Mathematical truth, for Cauchy, was an "inestimable treasure" of "divine beauty" and should therefore be cherished for its own sake. The practical utility of mathematics for engineering and technology is certainly admirable and desirable, but it is not the reason that mathematics should be studied. The road to "heavenly bliss" goes not through utility but through the pursuit of mathematical "immortal truth" for its own sake. For Cauchy, mathematical truth, with all its "celestial charms," stands alone, pure in its splendid isolation.

To attain this pristine mathematical reality, Cauchy argued, one must follow a single guiding principle: exactitude. Intuition and analogies based on physical reality are not nearly precise enough for proper mathematical reasoning. If recklessly pursued, they will inevitably lead to error. Instead, one must start anew and with care and precision define every term used and the exact range over which every statement is true.

Only by following these careful steps can one hope to attain the "divine beauty" of mathematical truth. Compromising on these principles meant, for Cauchy, compromising on the truth itself, and this betrayal was unthinkable. To his contemporaries, Cauchy may have seemed intransigent and stubborn in defense of extreme and unpopular positions; but in his own eyes he was only the meek bearer of the torch of truth in a world filled with error and darkness.

Cauchy's actual mathematical practice, as exemplified in the *Cours d'analyse* of 1821, follows these maxims to the letter. More than any of his contemporaries, Cauchy insisted that mathematics was its own unique reality, separate from the physical and social realities we know through our senses. Going against the grain of prevailing fashion, as well as the active opposition of his colleagues in the academy and the École polytechnique, he insisted that mathematics should be studied with exactitude and for its own sake, beginning with first principles. The end result was a new type of mathematical practice that over the following two centuries came to dominate the academic field of mathematics.

Both Cauchy's self-presentation as a martyr to truth and his mathematical creed are oddly reminiscent of his young contemporary, Galois. There is an irony in this, for not only were the political convictions of the two diametrically opposed, but Galois also viewed Cauchy as one of his chief tormentors. Nevertheless, the similarities between the two are striking. Cauchy believed that mathematics was a pure realm unto itself, characterized by logical rigor and precision; Galois, as we shall see, shared this belief and practiced it in his work. Cauchy believed that he was being persecuted by a wicked and uncomprehending world for his championship of the truth; Galois believed much the same. Cauchy's intransigence effectively isolated and marginalized him within the French scientific community; Galois' intransigence, as well as his deliberate provocations, had a similar—though more extreme—effect on his life and career. Both Galois and Cauchy believed that they were the bearers of eternal truth and were willing to suffer the indignities inflicted upon them by an uncomprehending world. Betraying the truth was unthinkable to them, and the price, any price, must be paid willingly. Such is the way of the martyr.

III

ROMANTIC MATHEMATICS

6

THE POETRY OF MATHEMATICS

The Romantics

The young revolutionary was going through the papers of his martyred friend when he came across a short verse:

> L'éternel cyprès m'environne;
> Plus pâle que la pâle automne,
> Je m'incline vers le tombeau.
>
> (The eternal cypress surrounds me;
> Paler than the pallor of autumn,
> I lean down toward the grave.)

Haunting lines they were, and prophetic, for their author died shortly after penning them. In the course of a stormy political career on the radical fringes, the young poet was ensnared in a love affair and was shot to death in a duel on a cold Parisian dawn. His last words were directed at his younger brother, who was overcome with grief seeing him on his deathbed: "Don't cry. I need all my courage to die at twenty."

The tragic death of this young poetic soul was a stunning blow to his friends and fellow radicals. Thousands showed up at his funeral to hail him as a hero of the revolution and bemoan the fate that had overtaken the most tender heart and most brilliant mind among them. In radical journals and political gatherings his friends railed against the pettiness of men who condemned this tender genius to obscurity, and the harshness of a world that found no place in its bosom for one of its most gifted sons. He was, of course, Évariste Galois—radical revolutionary, mathematical prodigy, and, it seems, aspiring romantic poet. The man who discovered the poem was his close friend and comrade in arms, Auguste Chevalier.[1] Devastating as Galois' tragic fate was to his acquaintances, it must nevertheless be admitted that it was not altogether unusual, but rather emblematic of artistic souls in its time. For in the early nineteenth century a tragic life and an early and sudden death were viewed not as senseless accidents but rather as essential components of the life stories of the greatest poets, artists, and musicians in the age known as "High Romanticism."

Romanticism, like *Renaissance* or the *scientific revolution*, is one of those terms that are often criticized for their vagueness and multiplicity of meanings but are nevertheless defiantly useful for our historical understanding. Even without a precise definition—or perhaps because of it—Romanticism refers to a cultural attitude clearly recognizable in the early decades of the nineteenth century. If there was a core attitude shared by the romantics of all stripes, it was their disillusionment with the ideals of the Enlightenment. The great thinkers of the eighteenth century Enlightenment, from Voltaire to d'Alembert to Condorcet, possessed an unbounded faith in the power of reason to understand the world and guide human action. The precise nature of rationality was certainly open to debate, and as we have seen, even close friends such as d'Alembert and Diderot could disagree profoundly on this issue. But they all agreed that reason was the key to human progress and betterment, and they fought to clear its path by sweeping aside irrational "superstition" inherited from the benighted past.[2]

The political turmoil of the final decades of the eighteenth century shook this optimistic faith to the core. The French Revolution, born of Enlightenment ideals of reason and reform, had unleashed psychic forces that the philosophes had never imagined. Progressive and reasonable demands by the populace in the early stages of the Revolution were soon followed by mob rule, riots, war, and terror. The Enlightenment faith in the unstoppable march of reason suddenly appeared misguided and hopelessly naïve in the face of the madness that had taken over the world. Romanticism was, first and foremost, a cultural manifestation of the disillusionment with Enlightenment rationalism.[3]

As a self-aware cultural movement, Romanticism spanned the range of fields from poetry to painting, music, sculpture, and architecture. The meaning of the term varied greatly between these fields, but all romantics shared a deep skepticism of reason as a way to understand and control the world. Instead of interpreting the world through the power of rationality, they sought to engage it on a deeper, emotional level or rise above it to higher and more perfect realms. The poetry of Wordsworth and Keats, the music of Beethoven and Chopin, and the paintings of Caspar David Friedrich and Théodore Géricault expressed a profound inner striving and appealed to something deeper in the soul than cold rational analysis. Skeptical of the power of reason to penetrate to the true essence of reality, they sought an unmediated connection with an untamable world. They were embarked on an unceasing quest for the sublime, a quality in the world that transcended both sense perception and rational analysis and was more fundamental, and more real, than either.

The quest for the sublime was inextricably bound with the persona of the romantic seeker, he who breaks the bonds of our mundane world and reaches out to higher spheres, inaccessible to most men. He is a young man (for he is almost always a man), pure of heart and restless of soul, who cannot help but search out the sublime truths that lie beyond the commonplaces he finds around him. Although he lives in our familiar world, his dark, piercing gaze is focused elsewhere and sets him apart from other men. He is clearly superior to his fellows, but his life is a hard one, for he is never truly at home in his commonplace surroundings. All too often he is misunderstood, his genius unrecognized and unrewarded by the lesser men who hold the reins of power. A beautiful misfit in a coarse and imperfect world, the romantic hero is a doomed soul whose quest for the sublime leads to loneliness, alienation, and all too often an early death. But in the few years allotted to him, the romantic hero burns more fiercely and shines brighter than any of his fellows, however long lived.

Lord Byron may have been the most famous of the tragic heroes of Romanticism. Having been chased out of London for living outside the social conventions of his time, he died in 1824 at age 36 fighting to free the Greeks from despotic Ottoman rule. The death of his friend Percy Bysshe Shelley by drowning two years earlier was just as tragic, if less heroic, as was the untimely death of John Keats, who perished of tuberculosis in 1821 at the age of 25. The German poet and philosopher Novalis died at the tender age of 28, while his contemporary Friedrich Hölderlin lived to a ripe old 73 but went insane and lived the last 36 years of his life effectively a prisoner in a friend's house.

The great musicians of that era seemed equally victimized by the Fates. Beethoven suffered the ultimate torment of a composer by gradually going deaf, and Chopin died at 39, at the height of his success, of tuberculosis. Mozart, who died at age 35 in 1791, may have belonged to an earlier generation, but he was warmly embraced by the romantics, who saw him as one of their own. To them, his untimely death and burial in an unmarked pauper's grave were emblematic of the tragic fate of a genius who remained true to his artistic calling. In the age of High Romanticism, a tormented life followed by an untimely death came to be seen as a good predictor for true genius trying to make its way in the world.[4]

Mathematical Poets

Galois, however, the young man who felt surrounded by the "eternal cypress" and perished at 20, was neither a poet, a musician, nor an artist in any conventional sense of the term, but a mathematician. His lasting fame rests not on poetic verses that tug at the heartstrings, nor on sublime symphonies performed before packed houses, but rather on his contributions to an esoteric field of high mathematics that even today can be appreciated only by a select and highly trained few. Misunderstood and marginalized by his contemporaries and cut down in the flower of his youth and promise, Galois is a paradigmatic romantic hero. His simple poem with its premonition of death suggests that he viewed himself in this light and perhaps acted in ways to bring about the expected end of his life's drama. His legend as it circulated and grew in the decades after his death shows that he was understood in precisely this way, first by his friends, later by fellow mathematicians and scientists, and ultimately by substantial segments of the public at large. As disseminated by Dupuy, Sarton, and Bell, the legend of Galois is of a piece with the legends of the romantic poets and artists of his time, although arguably it was more dramatic than any of them. In the romantic imagination the untimely deaths of Byron and Shelley, Novalis and Chopin, were inseparable from their spark of genius, which burned fiercely and consumed them while still in the prime of youth. Their tragic end was not seen merely as the conclusion of a productive career but as an essential component of their lives, the inevitable result of their creative fires.

"Even though it encounter no malignant enemy from without, genius will be sure to find within itself an enemy ready to bring calamity upon it," wrote the German romantic poet Heinrich Heine (1797–1856). "This is why the history of great men is always a martyrology: . . . they suffer for their own greatness, for the grand manner of their being, for their hatred of philistinism, for the discomfort they feel among the pretentious commonplaces, the mean trivialities of their surroundings—a discomfort that readily leads them to extravagances."[5] Intended as an insight into the restless souls and tragic fates of artists, poets, and musicians, Heine's comment might as well be a meditation on the life of the mathematician Galois. Frustrated at having to deal with the mediocrities who ruled his world—from his teachers and the examiners of the École polytechnique to the grandees of the academy—Galois was indeed driven to extravagances in both his political and personal life. And no one more than Galois can be said to have found his true enemy in himself.

So it was that as Galois' fame grew in the decades after his death, he came to be viewed as a pure romantic hero, alienated from his mediocre contemporaries and a victim of his own devouring genius. "When one sees how terribly fast this ardent soul, this wretched tormented heart were consumed, one can but think of the beautiful meteoric showers of a summer night," wrote George Sarton, the most poetic of Galois' biographers.[6] "If Galois had been simply a mathematician of considerable ability, his life would have been far less tragic . . . ; instead of which, the furor of mathematics . . . possessed him and he had no alternative but absolute surrender to his destiny."[7] The Galois of legend burned fiercely, exhausting his life in a single meteoric burst of unparalleled creativity. The inner flame that was the source of his genius devoured Galois, just as the creative fires manifest in the works of Byron, Shelley, and Novalis could not but extinguish the lives of these poetic souls at a young age.

As we have seen, Galois was far from the only mathematician of his age to be celebrated as a tragic romantic hero. Niels Henrik Abel was another, whose brilliant contributions to mathematics were cut short by his death at age 26 from the most romantic of illnesses, tuberculosis. Although the amiable and unassuming Abel never viewed himself in such a dramatic light, that made little difference to his lasting reputation. From the moment of his death onward Abel was hailed as a tragic romantic soul whose innocence made him the target of petty men and whose frail body wilted under the strain of his all-consuming mathematical genius. Even Augustin-Louis Cauchy, as much reviled for his arrogance and rigidity as he was admired for his mathematical acumen, viewed himself as a persecuted martyr, hounded by lesser men for standing up for what was true and holy. Although he was widely regarded as the chief villain in the legends of Galois and Abel, in his own eyes and those of his admirers he was much like them—a misunderstood romantic genius at odds with a hard-hearted and jaded world. A true artist "must keep faith in himself while the incredulous world assails him with its utter disbelief, he must stand up against mankind and be his sole disciple," wrote the American romantic Nathaniel Hawthorne in words that would have rung true to Galois, Abel, and Cauchy.[8] For each in his own way viewed himself and was viewed by others as a misunderstood genius, at odds with uncomprehending and uncaring mankind.

The life stories of Galois, Abel, and Cauchy were disseminated widely, acquiring certain mythical elements along the way that did not square well with historical fact. Galois was said to have written his entire mathematical testament in a letter on the eve of his duel, whereas in fact his mathematical contributions were contained not in his "last letter" but in his memoir to the academy and his published articles. Abel was transformed from the sociable young man he was in life to a tormented loner, and Cauchy remained as controversial in death as he was in life—the purest of martyrs to his ultra admirers, a scheming villain to his radical enemies. Both views are questionable, but such inaccuracies mattered not at all. The stories, after all, are not meant to provide an accurate historical account but a moral lesson. To mathematicians and the broader public alike, they represent an image of what a true mathematician is: a tragic romantic hero in the mold of a Byron, Shelley, or Van Gogh.

The famous tales of Galois and his contemporaries marked the emergence of a new understanding of what a mathematician is that contrasted sharply with older views. In the eighteenth century, as we have seen, the mathematical universe was ruled by the likes of Maupertuis, d'Alembert, and Euler, worldly and successful men who occupied the most prestigious academic chairs in Europe. To be a grand géomètre in the eighteenth century meant being courted by kings and emperors and pronouncing with authority on the cultural and philosophical issues of the day. But quite suddenly, beginning in the early nineteenth century, the meaning of being a mathematician and the images and stories associated with the role underwent a profound transformation. The true mathematician now became a lonely and tormented genius, unacknowledged by his contemporaries, who burned brightly for a while before being consumed by his own inner fire. The "worldly geometer" of the Enlightenment had been transformed into a romantic "mathematical poet."

This is not to say that the majority of mathematicians of that age or since fit the mold of the tragic genius. Certainly most mathematicians led lives far less dramatic than Galois, and some of the most respected, such as Karl Weierstrass (1815–1897) and David Hilbert, enjoyed the highest academic distinctions and lived to a ripe old age. Both were highly influential mathematicians whose research programs structured much of the work in the field in the nineteenth and twentieth centuries. But their fame extended no further than their mathematical colleagues and was based solely on their mathematical prowess. Only those with a particular interest in mathematical biographies took any interest in their relatively unremarkable life stories.

The case was very different for Galois and Abel, for their stories were the embodiment of a new mathematical persona that emerged in the early nineteenth century and was inextricably linked to tales of alienation and tragedy.[9] These romantic martyrs became icons to practicing mathematicians, as well the broader public, their legends a stark representation of what it means to be a mathematician. Their stories were disseminated widely, but also deeply—passed on by teachers and professors to young, aspiring mathematicians. Surely it is no coincidence that each new generation of mathematicians since then has produced its own mathematical poets, individuals as famous for their tragic lives as for their mathematical genius. These are mathematicians whose biographies follow the broad outlines of the martyr legend, as established in the generation of Galois and Abel. Their fame and the broad appeal of their stories remind us that the age of the romantic mathematical martyr, born in the turmoil of the early nineteenth century, has not come to a close.

Bernhard Riemann (1826–1866) was widely acknowledged as the most brilliant and versatile mathematician of his age and played a central role in shifting the center of mathematical research from France to Germany.[10] Shortly after his marriage in 1862 he contracted consumption, later known as pulmonary tuberculosis, the malady that had already claimed Abel before him. Despite repeated convalescent trips to Italy, he died of the disease at the age of 39, leaving behind masses of unpublished notes. Legend has it that Riemann's housekeeper burned many of his papers while tidying up his rooms after his death, before his colleagues at the University of Göttingen stepped in to save what was left. What inestimable treasures that overzealous housekeeper cost the world of mathematics will never be known.

Georg Cantor was the discoverer of transfinite numbers and the founder of set theory, a mathematical system that interprets all numbers (integers, rationals, reals, and so on) as infinite sets of different powers and is considered the conceptual foundation for modern mathematics. Unfortunately, the enmity of the well-connected Leopold Kronecker (1823–1891), who was the editor of Crelle's journal at the time, doomed him to a career at the University of Halle, on the margins of the mathematical world. There he suffered repeated mental breakdowns and ultimately died as a patient in an insane asylum. It is, of course, impossible to say whether Cantor's mental illness was related to his career troubles or his mathematical research, but legend has often linked his madness to both. The Cantor of legend is often depicted as a modern Faust, driven mad by his immersion in the most abstract and paradoxical branches of mathematics.

Srinivasa Ramanujan was a self-taught mathematical prodigy from Kumbakonam in India, possessed of an uncanny intuitive understanding of numbers and their hidden interrelationships. In 1913 he was working as a clerk in the Imperial service in Madras when, seeking recognition for his work, he wrote to G. H. Hardy, the leading British mathematician of the day. Hardy was so impressed by Ramanujan's original results that he arranged for the young Indian to come to Cambridge and work with him. Ramanujan's subsequent five-year stay in England proved both triumphant and tragic. He published numerous papers of great depth and originality and, in spite of his lack of formal credentials, was elected a fellow of the Royal Society. But he was also stranded far from home by the Great War raging in Europe, suffered bouts of depression, and ultimately fell victim to a mysterious illness that kept him hospitalized for months at a time. After the war he managed to return home to India, if only to spend his final days with his family. He died there a few months after his arrival at the age of 33.[11]

The list goes on. Kurt Gödel was a young Austrian logician in the 1920s who sat quietly in the wings during the philosophical gatherings of the celebrated "Vienna Circle." In 1930 he startled his colleagues by announcing that he had proven that any complex formal system, such as mathematics, contains certain statements that are true in themselves but cannot be proven. This remarkable result undermined some of the core assumptions of the Vienna Circle's logical positivism, which viewed mathematics as a shorthand for expressing rigorous logical relations. It also challenged Bertrand Russell's efforts to reduce all of mathematics to a systematic logical edifice, and David Hilbert's "formalism," which posited that mathematical statements were true if and only if they were correctly derived from other mathematical statements.

Forced to flee his home in 1939 by the threat of recruitment to the German army, Gödel ended up in Princeton as a fellow at the Institute for Advanced Study. Although he spent more than half his life there, Gödel was forever an alien in suburban New Jersey. Often at odds with his fellows at the institute in the early years, he became increasingly paranoid over time, convinced that his food was being poisoned by unnamed villains. When his wife fell ill and could no longer cook for him, he refused to eat and starved himself to death at a local hospital. Like Cantor's madness, Gödel's mental illness has often been linked to his mathematical work and his breathtakingly paradoxical results. The moral is clear: he who ventures out to the frontiers of logic and reason does so at his own risk, for he may never find his way back from that forbidden land.[12]

As a young mathematician in Cambridge, Alan Turing (1912–1954) followed up on Gödel's and Hilbert's work in mathematical logic and laid the conceptual groundwork for the future development of computers. During World War II he was recruited to the Code and Cypher School at Bletchley Park, where he contributed decisively to breaking the code of the German Enigma machine. At the end of the war he was inducted into the Order of the British Empire (OBE) for his vital contributions to the war effort. In the following years Turing continued his groundbreaking work in computer science, but the government that he had served so faithfully turned against him. In 1952 he was prosecuted for homosexuality and was given a series of estrogen shots in lieu of prison time. His security clearance was revoked, preventing him from continuing his code-breaking work, and he was stalked by the clandestine services, who suddenly considered him a dire security risk. He died mysteriously at age 42 in his laboratory after taking a bite from an apple coated with potassium cyanide. Although suicide was the most likely cause, darker theories regarding the circumstances of his death have persisted ever since.

Perhaps the most famous tragic mathematician today is John Nash, hero of the best-selling biography *A Beautiful Mind* and the feature film based on it.[13] "Nash's genius was of that mysterious variety more often associated with music and art than with the oldest of all sciences," wrote his biographer, Sylvia Nasar, and surely it is no coincidence that in her telling Nash suffers the fate of true artists and musicians.[14] Nash's mathematical brilliance, according to Nasar, is inseparable from his eccentricity and mental illness. Like Cantor and Gödel before him, but also like Beethoven and Van Gogh, he had ventured too far from our familiar world into faraway realms of purity and beauty. Although over decades he did manage to find his way back to sanity, the experience forever set him apart from other men.

Equally eccentric is Nash's exact contemporary, Alexander Grothendieck, 1966 recipient of the Fields Medal, the highest professional honor awarded to mathematicians, and one of the most influential mathematicians of the twentieth century. Grothendieck was a peace activist who at the height of his mathematical career took the time to teach advanced mathematics in the forests around Hanoi as the city was being bombed by American B-52s. In the 1970s and 1980s he grew increasingly critical of the close relationship between academic institutions and the military and gradually withdrew from his academic positions. He stopped publishing in mainstream scholarly journals and finally, in 1991, left his home in Montpellier and disappeared. He is rumored to be living in southern France or the Pyrenees, receiving no visitors.

A Modern Galois

A particularly striking example of the persistence of the legend of the romantic mathematician appeared in the August 28, 2006, issue of the *New Yorker,* which carried a story titled "Manifold Destiny" about trouble brewing in the international mathematical community.[15] It told the story of Grigory Perelman, a brilliant Russian mathematician who had taken on the most famous unsolved problem in modern mathematics: the Poincaré conjecture. At the dawn of the twentieth century the French mathematician Henri Poincaré suggested that all closed, simply connected, three-dimensional manifolds are, topologically speaking, spheres. Although it was universally believed to be true, and although similar theorems had been proven for dimensions greater than three, the Poincaré conjecture had resisted all attempts at proof for nearly a hundred years.

Over eight months, beginning in November 2002, Perelman posted a proof of the conjecture in regular installments on a mathematics website. His colleagues around the world followed his postings with increasing excitement as each segment of the proof was worked over, clarified, and verified by experts in the field. By the time Perelman had posted the last segment, a growing consensus had emerged: Perelman's proof was correct, and the Poincaré conjecture was no longer a conjecture: it was now a proven theorem.

Like other great mathematicians of the past, Grigory Perelman is an eccentric. Having spent years in the United States and widely recognized for his mathematical genius, he was offered his choice of positions in the top academic departments in the country. But Perelman declined them all and returned instead to St. Petersburg, where he moved in with his mother. There, working on the staff of a local mathematics institute and living ascetically on a salary of less than $100 a month, Perelman devoted his time to the Poincaré conjecture. In the spring of 2003, as the mathematical community was abuzz with the news that the century-old problem had been solved, Perelman went on a monthlong lecture tour in the United States, speaking of his work before audiences of leading mathematicians. He then returned to St. Petersburg and made no further effort to publicize, or even publish, his proof. His work, he believed, spoke for itself.

According to the *New Yorker* article, however, not everyone was pleased with Perelman's achievement. Shing-Tung Yau is a Chinese American mathematician at Harvard, a former winner of the Fields Medal and widely viewed as one of the leading mathematicians alive. He is also one of the most powerful: over the years Yau has taught generations of students and has helped place the brightest of them in some of the most prestigious academic institutions in the world. He is particularly well connected in China, where he is the director of mathematical institutes in Beijing and Hong Kong and is a personal friend of former Chinese president Jiang Zemin. His goal, the article's authors state, is to be recognized as the leader of Chinese mathematics.

Yau may well be an honorable man whose only goal is to promote mathematical knowledge in his native China and around the world; that, however, is not the story presented in the *New Yorker.* Yau, the article claims, had been working on the Poincaré conjecture for some years and was not pleased that Perelman had upstaged him. He quickly arranged for two of his former students to write up a proof of the conjecture that, he claimed, filled in critical "gaps" in Perelman's work. The new proof was not subjected to peer review and was published with great fanfare in 2006 in the *Asian Journal of Mathematics,* which Yau edits. Yau then publicly "presented" the proof to famed physicist Stephen Hawking, who was in Beijing as the keynote speaker at a conference Yau had organized. Not so subtly, according to the *New Yorker,* Yau was moving to steal the honors rightfully due to Perelman and claim them instead for himself, his students, and Chinese mathematics.

To many of his colleagues, these actions seemed like a serious breach of protocol. Here was Yau, a powerful institutional mathematician, using his influence and connections to disown the eccentric recluse Perelman. "Politics, power, and control have no legitimate role in our community, and they threaten the integrity of our field," said mathematician Phillip Griffiths, former director of the Institute for Advanced Studies in Princeton.[16] Perelman too was dismayed. He quit his position in St. Petersburg and retired from mathematics in protest over the field's "lax ethics." When the president of the International Mathematical Union visited him to urge him to accept a Fields Medal for his accomplishment, he flatly turned him down. "It is not people who break ethical standards who are regarded as aliens," he complained to the authors of the *New Yorker* article. "It is people like me who are isolated." Galois, railing in Sainte-Pélagie against the "important men of science," could not have said it better.

One cannot help but recognize the outlines of the Galois legend in the story of Perelman and Yau, as reported in the *New Yorker.* Once again an innocent young genius in pursuit of true knowledge presents his masterpiece to the mathematical community; and once again his efforts are stymied by a power-hungry established mathematician intent on robbing him of the recognition that is his due. Shocked by the betrayal of those he considered the guardians of truth, the young genius retreats to the margins of society and retires from mathematics, never to work in the field again. Perelman in the story makes for a plausible, if less volatile, Galois. Yau is here the spitting image of Cauchy as viewed through the prism of the Galois legend: a world-famous mathematician, a hoarder of honors and positions, indifferent to the struggles of a young genius and interested only in garnering more laurels for himself.

It should be noted that the true situations of Galois and Perelman were, in fact, very different. Unlike the desperate French firebrand, the young Russian was widely recognized for his accomplishment, was offered a choice of prestigious academic positions, and was awarded the highest honor for a mathematician, a Fields Medal. His decision to refuse them all can hardly be laid at the doorstep of the mathematical community. It is striking, therefore, that the *New Yorker* article's authors, several leading mathematicians whom they quote, and, apparently, Perelman himself view the matter in precisely Galois-like terms: an innocent mathematical genius driven from the field by a powerful establishment mathematician. One hundred seventy years after his death Galois' legend still holds sway in academic mathematics departments, his short life a model for the pure and true mathematical life.

"A Maker of Patterns"

One of the most compelling statements of the parallels between mathematicians and artists comes from none other than Ramanujan's host and mentor, G. H. Hardy. Although Hardy was undoubtedly one of the leading mathematicians of his generation, with contributions ranging from analysis to number theory, he is best known to the broader public as the author of *A Mathematician's Apology,* a beautifully written and witty exposition of a mathematician's understanding of his field.[17] Much of this immensely popular work, first published in 1940, is devoted to arguing that the similarities between art and mathematics are not coincidental but reflect deep affinities between the fields. Unlike the natural sciences, Hardy argues, the purpose of doing mathematics is not to attain useful results with practical applications. To the contrary, truly interesting mathematics is almost always useless and is pursued solely for the sake of truth and beauty.

"A mathematician, like a painter or a poet, is a maker of patterns," Hardy writes, but his art is of a special kind: "If his patterns are more enduring than theirs, it is because they are made with ideas."[18] Nor do the commonalities between art and mathematics end there: "The mathematician's patterns, like the painter's or the poet's, must be *beautiful;* the ideas, like the colours or the words, must fit together in a harmonious way. Beauty is the first test: there is no place in the world for ugly mathematics."[19] He then concludes: "Real mathematics must be justified as art if it can be justified at all."[20]

For Hardy, at least, "real" mathematicians are indeed artists, creating lasting and beautiful patterns in the manner of poets and painters. Unlike natural scientists and engineers, whose work is judged by its utility and its correspondence to the physical world, mathematics, like art, is judged in terms of its beauty and internal harmony alone. Hardy did not deny, of course, that mathematics can be extremely useful in the pursuit of practical goals. This, however, was boring and "trivial" mathematics, and he had nothing but pity for applied mathematicians who had to focus their energies on the ugliest and least interesting branches of their field. Applied mathematicians, for Hardy, may indeed have much in common with physical scientists who investigate the world as it is, and with engineers who build things for power and gain. But "real" mathematicians are artists whose sole interest is the sublime beauty of the patterns they create with their minds.

If mathematics is a form of art or poetry, and mathematicians are artists and poets, it is hardly surprising that the lives of mathematicians would follow the romantic trajectory of artists' lives. So it was, as we have seen, for mathematicians from the time of Abel and Galois to that of Nash and Grothendieck, who saw themselves, or were viewed by others, as tragic heroes in pursuit of sublime truth. Hardy himself, it should be noted, was an extremely successful academic who at different times held the top mathematical posts at both Oxford and Cambridge and died at the respectable age of 70. As such, he would not normally be considered a tragic figure, but it is significant that others, nevertheless, saw him as such. To his friend C. P. Snow, *A Mathematician's Apology,* written when Hardy was in his sixties, is "a passionate lament for creative powers that used to be and will never come again."[21] Hardy, according to Snow, lost his will to live once he realized that his access to the mathematical world of truth and beauty was forever closed to him, and he was to live out his days in our mundane world. Hardy's death in 1947, Snow writes, while not technically a suicide, might as well have been one. To his friend, at least, even the highly successful Hardy was ultimately a tragic romantic hero, reaching out to purity and truth, but ultimately trapped in a world of trivialities.

From Galois to Ramanujan to Nash, it appears, the brightest mathematical lights were viewed as tragic romantic figures, and Hardy's *Apology* does much to explain why that is so. Mathematics, in Hardy's view, is a creative art, and a mathematician, accordingly, is an artist, living his life in a quest for the sublime. Furthermore, Hardy insisted repeatedly in his book, this view of the mathematician as an artist is not unusual and not his alone, but is representative of prevailing attitudes among his mathematical peers. No less than the public at large, it seems, some mathematicians, at least, saw themselves as romantic artists on a doomed quest.

The contrast between the mathematical stories that appeared in the nineteenth century and the narratives that prevailed only a generation before are profound. There were no mathematical poets in the eighteenth century, no Galois or Abel, no Ramanujan or Nash, and much the same can be said about the seventeenth, sixteenth, and earlier centuries, going back to medieval times. This, of course, is not to say that mathematicians had not suffered misfortune or abuse at the hand of their contemporaries. Some aspiring mathematicians of that age were undoubtedly failures, and some were undoubtedly persecuted by their more powerful colleagues. One might think, for example, of Johann Samuel König (1712–1757), who was, perhaps, a mathematician in the mold of Galois. Although lacking in institutional support, he publicly took on two of the most powerful geometers in Europe—Maupertuis, who was president of the Berlin Academy, and his friend Euler, the leading mathematician in Europe. At issue was Maupertuis' principle of least action, which he viewed as the crowing achievement of his career, but which König claimed had been discovered previously by Leibniz.

König considered himself a disciple of Leibniz, which explains why he championed his case for priority. But it hardly explains why König publicly challenged Maupertuis, who by all accounts had previously been friendly toward him and was even something of a patron. Only a Galois-like pigheadedness on König's part and a talent for turning friends into enemies can account for that. Predictably, Maupertuis was outraged at what he considered a betrayal and a slight to his greatest accomplishment, and in the long controversy and investigations that followed, König's reputation was effectively destroyed. He died an outcast from the Republic of Letters at the age of 45.

Had König been born in 1812 rather than 1712, his stormy career might have placed him in the ranks of those brilliant mathematicians who stood up for the truth in the face of overwhelming odds. Like Galois and Cauchy, he did not hesitate to challenge the scientific establishment in the name of what he believed was right, and like them, he suffered severe professional and personal consequences. To the romantic eye of the nineteenth century, such martyrdom was a strong indicator of profound mathematical insight, and one can well imagine that had König been a child of the nineteenth century, he might be remembered today as another genius crushed by institutional indifference.

The verdict of the eighteenth century, however, was different: König was marginalized and his mathematical skill discredited. He was ostracized from the Republic of Letters, and that was that—he was doomed to oblivion, and neither he nor his reputation could be revived. "His book on dynamics is buried with him, if it ever existed," the academician Jean-Bertrand Merian wrote to Maupertuis with satisfaction when news came of König's death in 1757.[22] And so it proved. König was excluded from the ranks of legitimate mathematicians and from his day to ours has remained largely forgotten. There was no glory in the eighteenth century for mathematicians like König who lived short and tragic lives. The iconic geometers of the age were men like d'Alembert and Euler, not Galois and

Abel, successful men of affairs, not alienated loners. Sad mathematicians like König might be pitied, but they were certainly not admired for their misfortunes.

Why Natural Scientists Are Not Poetic

The paradigmatic tragic mathematician, then, was a novelty when he appeared on the scene in the early nineteenth century, a mathematical persona very much at odds with earlier conceptions. In their own eyes and the eyes of their contemporaries, nineteenth-century mathematicians were very different people than their predecessors in the eighteenth century and earlier had been. Even so, one might still wonder whether the shift in the prevailing narrative was unique to mathematics or was shared with the other sciences. It is possible, after all, that just as mathematical imagery absorbed some of the dominant themes of High Romanticism, redefining what it meant to be a mathematician, so did other sciences. Did idealized romantic physicists and naturalists capture the imagination of their contemporaries at the same time that Galois and Abel became paradigmatic examples of the mathematical life?

If that were the case, then one might conclude that the transformation of the mathematical narrative in the early nineteenth century had little to do with the practice of mathematics itself and was part of a broader phenomenon affecting the sciences. Scientists and mathematicians, one might argue, acquired the characteristics of tragic heroes simply because that was the fashion of that romantic age. But if the new imagery infected such a broad range of activities, it would follow that the romantic hero theme is unrelated to any specific developments in each separate field. In other words, if poetic figures like Galois are as omnipresent among contemporary physicists, chemists, and engineers as they are among mathematicians, then the emergence of this new romantic persona can teach us little about the developments taking place within the field of mathematics.

This, however, was not the case. The leading natural scientists of the nineteenth century were not a romantic lot, nor were they inclined toward the tragic or poetic. To the contrary, they were practical and energetic men who, like their predecessors in the eighteenth century, held positions of power and distinction and were involved in a broad range of ventures. William Thomson was perhaps emblematic of this breed of natural scientists, although he was exceptionally successful. He not only developed the physical theories of electricity and thermodynamics but also became a world-famous engineer responsible for laying the first transatlantic underwater cable. For his accomplishments he was created Baron Kelvin of Largs and served three terms as president of the Royal Society, including the last 12 years of his life. Thomson's German friend, Hermann von Helmholtz (1821–1894), was another of these intrepid natural scientists. Trained as a physician, he made crucial contributions to physiology and the theory of vision before turning his attention to electrodynamics and thermodynamics. Throughout his long career he held professorships at several of the leading German universities before being granted his own physics institute at the University of Berlin. His popular writings on topics related to the philosophy of science enjoyed enormous popularity and made Helmholtz a public spokesman for the natural sciences.

Needless to say, not all leading nineteenth-century physicists enjoyed such long and prosperous careers, just as not all mathematical lights suffered the fate of Galois or Abel. James Clerk Maxwell (1831–1879), for example, famous as the author of the Maxwell equations uniting the forces of electricity and magnetism, died of cancer at the age of 48. But tragic though his premature death was, it was not viewed as the inevitable final chapter of a tormented life, but as the sudden and untimely end of a brilliant career. At the time of his death Maxwell was the first Cavendish Professor of Physics at Cambridge, a member of the Royal Society, and the head of the Cavendish Laboratory. He was unquestionably an enormously successful member of the scientific community who had the misfortune of dying young. Although his untimely death was indeed a tragedy, there was nothing tragic about his life.

Ludwig Boltzmann, whose work contributed to the founding of statistical mechanics and atomic theory, was perhaps a sadder case, a victim of chronic depression who ended his life by suicide. But like Maxwell's, his life, although it ended in tragedy, was anything but tragic. He held professorships at some of the leading universities in Austria and Germany and was known as a spellbinding speaker whose lectures on the history and philosophy of science attracted enormous crowds. He was also happily married and a father of five, a role in life that is quite unimaginable for the romantic mathematical poets.

Finally, the best-known naturalist of the nineteenth century was likely Charles Darwin (1809–1882), whose life generated a great deal of interest in the broader public. As a young man Darwin traveled the world aboard

HMS *Beagle,* an experience he described in his popular *Journal and Remarks* (better known as *The Voyage of the Beagle*) in 1839. The journey established Darwin as a grand adventurer, as well as a rising star in the British scientific elite, but he did not stay long in London to enjoy his newfound success. Within a few years of his return he retired with his wife to the country estate of Down House, where, over the following decades, he fathered ten children, seven of whom survived to adulthood. He remained at Down House for the rest of his life, leaving it only on the rarest of occasions.

The publication of *On the Origin of Species* in 1859 brought Darwin sensational and unexpected fame, instantly transforming him into a heroic and wise patriarch to some and an apelike figure of fun for others, depending on their religious and political leanings. He remained at the center of the storm to his death, an unlikely combination of a perfect country gentleman and a radical intellectual revolutionary. Darwin was undoubtedly viewed as eccentric, and it is likely that his early retirement, large family, and bearded countenance that recalled an Old Testament prophet added significantly to his fame. But was he a tragic figure in the mold of romantic poets or mathematicians? Hardly. He traveled the world like Byron, but his journeys brought him success and the respect of his peers, not death. He withdrew to his country estate, but that did not mark a disillusionment with science work but the beginning of a new and dramatic chapter. In his life he may have suffered tragedy, most notably the death of his daughter Annie at age 10, but his public persona was that of a great and wise patriarch, not a tragic martyr.

Much the same can be said of the leading natural scientists of the twentieth century. Albert Einstein, the iconic scientist of the twentieth century, was undoubtedly viewed as eccentric, a small man with wild eyes and a lion's mane of white hair. It is probably no coincidence that one of his most famous portraits depicts him in the childish act of sticking his tongue out at the camera. Nevertheless, Einstein was by no means a tragic figure in the mold of a Galois, or even of his friend Kurt Gödel. In fact, it can be said that his popular depictions are rather similar to those of the "natural man" d'Alembert in the eighteenth century. Like d'Alembert, Einstein was viewed not as a romantic striver but as an eternal child who had never lost the innocence and curiosity of childhood and had stayed in tune with the rhythms of nature. His subject matter, like that of d'Alembert, was the real world that surrounds us, not Hardy's abstract universe of beauty and truth. Consequently, the secret of his genius was not that he had, like a mathematician, ascended to higher realms of truth and beauty, but rather that he had retained his simple childlike affinity with the physical world at an age when the rest of us had drawn away from it. Like d'Alembert, Einstein was perceived as a child of nature and was consequently always at home in our natural physical world.

All this, of course, is but a sampling of the most famous names in the natural sciences in the past two centuries, and it is not intended as a comprehensive survey of the lives and careers of leading scientists. The sample is sufficient, however, to establish this surprising fact: the romantic persona of the tragic hero, who strives toward beauty and truth only to be crushed by material circumstances, is completely absent from the natural sciences. As this limited survey makes clear, natural scientists are not romantic artists. It is safe to say that no leading natural scientist of the past two centuries, and no idealized biography of a natural scientist, has conformed to the iconic image of the tragic romantic hero. That particular image has been reserved exclusively for mathematicians.

The iconic tale of the tragic romantic mathematician, it seems clear, was both novel and exclusive in the early nineteenth century—novel because it ran counter to the images that had prevailed only a few years earlier, when mathematicians were more likely to be viewed as simple natural men than as striving romantic heroes; exclusive because the new story was reserved, among the sciences, to mathematics alone. Only mathematics came to be viewed as a quest for pure sublime truth, only mathematics was perceived as a creative art rather than a science, and only mathematicians became tragic romantic strivers in the manner of contemporary poets, painters, and musicians. It can well be said that in the early nineteenth century mathematics took leave of the natural sciences, which had been its companions for millennia, and sought a place for itself instead with the creative arts.

The Frenchman and the German

The shift in mathematics' place in the intellectual globe in the early nineteenth century is beautifully encapsulated in the exchange between Joseph Fourier and Carl Gustav Jacobi. These two leading mathematicians, one French, the other German, represented not only distinct generations but also competing national traditions. Fourier was a mathematician in the Enlightenment tradition, a practical man, and a direct heir of Maupertuis, d'Alembert, and Lagrange. A member of the 1798 expedition to Egypt and subsequently a high official under Napoleon, Fourier retained his influence during the Restoration and became the permanent secretary of the Academy of Sciences in 1822. Like his eighteenth-century predecessors, he strongly believed that mathematics was a science of the world, whose power lay in its ability to reveal the hidden patterns and harmonies that govern it but are normally hidden to the observer. Mathematics "brings all the most diverse phenomena together and discovers the hidden similarities which connect them," Fourier wrote in 1822. D'Alembert himself could not have said it better.[23]

Just as d'Alembert had focused on dynamics and Lagrange on a general theory of mechanics, Fourier concentrated his energies on another physical problem, the theory of heat propagation. His most important work was his *Théorie analytique de la chaleur* (Analytic Theory of Heat) of 1822, where he developed the trigonometric Fourier series for which he is best known today. In the introduction Fourier did not hide his suspicion of pure mathematics, practiced for its own sake. The methods developed in his book, he wrote, "lead to the last numerical calculations, and that one must demand of any investigation if one does not want to obtain merely useless transformations." His equations, he insisted, do not stand for themselves but for underlying relations that exist in the physical world: "The problem that we want to uncover lies no less hidden in the analytical formulas than in the physical problem itself."[24] Once again we are in the intellectual world of d'Alembert, who 70 years earlier had insisted that algebraic formulas are abstractions of physical relations.

Fourier was in his sixtieth year in 1828 when he was charged by the academy with reviewing the recent results obtained by the rising mathematical stars Abel and Jacobi on the theory of elliptic functions—a highly abstract study of a special class of complex functions. Fourier was clearly impressed with his young colleagues, reporting that their results belong to the most beautiful and deepest ones in mathematics. But true to his belief that mathematics was ultimately a natural science, and its object the physical world, Fourier could not refrain from a complaint: such highly qualified individuals should also direct their research interests to solving problems arising from the natural sciences, which, according to Fourier, are the ultimate measure of the progress of human intelligence. Abel and Jacobi's work might be beautiful and deep, Fourier argued, but it was also useless and hence, in the last account, pointless.[25]

By the time Jacobi heard of these remarks in 1830, a year had passed and both Fourier and Abel were dead. Jacobi, who was teaching at the University of Königsberg, had sent a copy of his book *Fundamenta nova theoriae functionum ellipticarum* to the Paris Academy, where Denis Poisson reported favorably on the work but also cited Fourier's remark the previous year. Jacobi was pleased with Poisson's review, writing to Legendre that he read it with "great pleasure," but he was clearly irked at Fourier's jibe that his work was useless. Fourier, he wrote to Legendre, was fundamentally mistaken, not just about Jacobi's and Abel's work but about the very nature of mathematics.

> It is true that in M. Fourier's opinion the principal aim of mathematics is its public utility and the explanation of natural phenomena; but a philosopher such as himself should have known that the only aim of science is the honor of the human spirit, and that under this title a question about numbers is worth as much as a question about the system of the world.[26]

Therein lay the difference between Fourier and Jacobi, between the French and German mathematical traditions, and, above all, between the older generation of Enlightenment mathematicians and the new generation of Jacobi, Abel, and Galois. For Fourier, mathematics is derived from the world, and its goal is to reveal the hidden structure of the world and describe it as it really is. Pure mathematics of the kind practiced by Abel and Jacobi is pointless because it is mired in empty, "useless transformations" that tell us nothing of the world and lead nowhere. Such mathematics might be beautiful and even deep, but it is ultimately a betrayal of what the field should be—a study of the physical world.

For Jacobi, in contrast, the physical structure of the world is entirely irrelevant for mathematics. Like Hardy a century later, he believed that mathematics, like art, is practiced for its own sake, for truth and beauty, whose pursuit is the highest calling of the human spirit. The subject matter of mathematics for Jacobi is not the physical world but mathematics itself, its perfect inner structure and the beautiful relationships between Platonic entities such as lines, planes, and numbers, known generally as mathematical objects. Jacobi strongly objected to Fourier's criticism that he and Abel had strayed too far from the natural world, and he took it on directly. "We are unhappy that most French geometers . . . have presently fallen into this error," he said in his 1832 inaugural lecture as ordinary professor at Königsberg. Making what would become a common distinction in the coming years between the French and German mathematical traditions, he added: "While they seek to obtain the only salvation of mathematics in physical problems, they desert that true and natural path of the discipline."[27]

All this does not mean that Jacobi considered mathematics irrelevant for the sciences or technology. To the contrary, Jacobi insisted that mathematics is an enormously powerful tool when it is applied to the tasks of describing the world and controlling it. His point was, rather, that mathematics does not depend on the natural world but on itself alone and must be held only to its pure internal standards. Whether a result was useful, Jacobi argued, was irrelevant to its value as a mathematical truth. And just as mathematics was independent of the physical world, so was the world independent of mathematics. Although mathematics could be effective in describing material reality, the correctness of mathematical transformations in itself tells us nothing about what is actually happening in the physical world. Reality, in the end, did not depend on correct mathematical reasoning.

No one was more conscious than Jacobi himself of the break he was making with the established tradition of Enlightenment mathematics. While professing great admiration for his predecessors, Jacobi took issue with their practices, and with those of Lagrange in particular. In his lectures on Lagrange's *Mécanique analytique* Jacobi warned his students not to be seduced by the power and elegance of Lagrange's system. For Lagrange, he argued, "Everything is reduced to mathematical operations . . . Nature is totally ignored."[28] Such breaching of the boundaries between mathematics and nature, he warned, can create serious confusion and lead to error. Physical laws, for Jacobi, cannot be proved mathematically, just as mathematics cannot be deduced from the physical world. The break with Enlightenment geometers, who considered the world mathematical and mathematics worldly, could not have been sharper.

The Land of Truth and Beauty

Jacobi was not the only mathematician of that age to insist that mathematics is the pursuit of truth and beauty and should be judged according to its rigorous inner standards. In 1828 Auguste Crelle, advising the Prussian authorities on the institution of the new Polytechnic Institute in Berlin, wrote that "mathematics should be explained in the first instance without regard for its applications, and without its being interrupted by them. It should develop purely from within itself and for itself, for only in this way can it be free to move and evolve in all directions."[29] Crelle's friend Abel also complained that the mathematicians of the older generation were far too focused on mathematical applications to pay heed to the proper standards of mathematics itself. "Poisson, Fourier, Ampère, etc. etc. are only occupied with magnetism and other physical things" and are not at all interested in pure mathematics, he wrote to Holmboe from Paris in 1826.[30] As a result, he warned, mathematics—and analysis in particular—had lost its way and was in danger of encountering insoluble paradoxes. "I will apply all my strength to bringing more light into the vast darkness that unquestionably exists in analysis," he wrote to Hansteen from Paris in 1826.

> It totally lacks any plan and system, so it is really very strange that it is studied by so many and worst of all, that it is not treated rigorously at all. There are very few theorems in the higher analysis which have been proved with convincing rigor . . . It is very strange that after such a procedure there exist only a few of the so-called paradoxes.[31]

The exception to French mathematicians' lamentable focus on physical problems, according to Abel, was Cauchy. "The excellent work by Cauchy, *Cours d'Analyse de l'école polytechnique,* which ought to be read by every analyst who loves rigor in mathematical investigations will serve as my guide," he announced proudly in the introduction to an article he published in Crelle's journal in 1826.[32] A few months later, in his letter to Holmboe from Paris, he explained his admiration for Cauchy: "Cauchy is fou [mad] and you can't get anywhere with him, although for the moment he is the mathematician who knows how mathematics must be done." "Cauchy is tremendously catholic and bigoted," he continued, "a very strange thing for a mathematician. Otherwise he is the only one working on pure mathematics." Mathematically, if not personally, Abel found in Cauchy a kindred spirit—one who valued mathematics for its pursuit of rigorous truth first, and practical applications only a distant second.

Jacobi, Crelle, Abel, and Cauchy were all members of a new generation of mathematicians who came into their own in the early decades of the nineteenth century, bringing with them a radical new perspective on the practice of mathematics. Whereas their Enlightenment predecessors viewed mathematics as an abstraction from the physical world, the new generation considered it an independent field concerned with pure mathematical objects and their interrelations. Whereas geometers from Johann Bernoulli to Fourier evaluated mathematics according to its utility and applicability to describing natural phenomena, the new mathematicians insisted that the field be evaluated purely in accordance with its own internal standards. And whereas Enlightenment mathematicians viewed the field as part of the natural sciences, for the new generation it was closer to the creative arts, a pure and sometimes tragic pursuit of the "inestimable treasure," truth.[33]

Simply put, whereas Enlightenment mathematics was concerned with this world, the new mathematics was focused on alternative universes, pure and beautiful, governed strictly by mathematical principles. Mathematicians, accordingly, are those who reach out to this better world and try to capture a glimpse of its glory. The most successful ones are those who have found their way to the mathematical realm and have observed its beauty and wonders firsthand. "Imaginary universes are so much more beautiful than the stupidly constructed 'real' one," wrote G. H. Hardy, and it is hardly surprising that those privileged few, who have attained these sublime heights, find it difficult to adjust to life in our mundane universe.[34] Like the captive in Plato's cave metaphor, who had seen the true glory of the forms, they can never again be content with life in the cave of shadows. They are the brightest mathematical lights, the Galois and Abels, the Cantors and Grothendiecks, and they are inevitably the ones who can never adjust to life in our own world. They live among us, but they ware not of us, and the price they pay for living their physical lives in our world is a heavy one—self-destructiveness, madness, and even early death. Their true home is elsewhere, their true lives are lived in a sublime and beautiful mathematical realm, and try as they might, they will never belong in our imperfect world, ruled by natural constraints and human folly. The fact that they must live their physical lives in such a world is, in a word, a tragedy.

The vision of mathematics as the study of a pure realm of truth and beauty is therefore inseparable from the image of the mathematician as a tragic romantic figure. Each of these requires the other: the romantic legend of the tragic mathematical artist implies that mathematics, like creative art, is a pursuit of a higher truth and sublime beauty. Hence mathematics is the study of a beautiful alternative world, studied, as Jacobi pronounced, for "the honor of the human spirit." At the same time, a conception of mathematics as insulated from any form of physical and human reality leads almost inevitably to the iconic view of the tragic mathematical genius, for in this view a mathematician is that blessed individual who is granted a special sight that allows him to observe, or even to reside within, the alternative mathematical universe. It is an inestimable gift, but it comes at a heavy price: striving for his true home in the realm of mathematics, but trapped in the harsh world of physical things and the petty company of lesser men, the mathematician is a tragic figure, doomed to a difficult and disastrous life. Inevitably, the paradigmatic new mathematician takes on the tragic and romantic outlines of a Galois, an Abel, or a Perelman.

7

PURITY AND RIGOR: THE BIRTH OF MODERN MATHEMATICS

Cauchy Reinvents the Calculus

What can one say of the "sublime mathematics" advocated by tragic poet-mathematicians? What does it consist of, and how does it differ from other approaches to the field? It is one thing to claim, as Jacobi did, that mathematics should be practiced "for the honor of the human spirit," or, like Hardy, that it is a form of creative art. It is quite another matter to create a body of mathematical work that reflects this vision of the field. We have already noted that Enlightenment geometers practiced a type of mathematics that flowed from their belief that algebraic expressions were an abstraction from true physical relations that prevailed in the world. It was a mathematical practice that disdained "pedantic" rigor and focused on physical problems drawn from the natural world. At the same time, as we have seen, Enlightenment mathematics gradually came to emphasize abstraction and generality, which were thought to represent the deep and hidden laws governing the world around us. But how would nineteenth-century mathematicians, who worked under a very different notion of what their field was about, practice their craft? In other words, what would a mathematics viewed by its practitioners as a sublime realm of truth and beauty look like?

A clue about the nature of this new mathematics can be found in Cauchy's introduction to his *Cours d'analyse* of 1821, where he berates his predecessors for the inexactitude and incertitude of their methods. In his own work, he proclaims, he will focus on the precise conditions in which each and every statement is true, and by doing so, he confidently asserts, "I make all incertitude disappear."[1] Cauchy then goes on to insist that mathematics should not transgress its bounds and impose its methods on other realms of knowledge:

> We are persuaded that there exist truths other than the truths of algebra, realities other than sensible objects. We cultivate the mathematical sciences with ardor, without wishing to extend them beyond their domain; and we will not imagine that one must approach history with formulas, nor support morals with theorems of algebra or the integral calculus.[2]

For the devout Cauchy, the "truths other than the truths of algebra" or sensible objects were undoubtedly a nod to the domain of revealed truth, which lay at the core of his Catholic faith. But there is a larger implication here: different types of truths are gained by different means and belong on their own domain. God's truth is known through revelation, material objects are known through the senses, and mathematical truths are known through mathematical reasoning alone. Any attempt to cross boundaries and mix one type of knowledge with another will result in confusion and error.

Cauchy is in effect taking a stand against Enlightenment geometers who viewed mathematics as an abstraction from physical reality, and the physical world as an expression of mathematical relations. Not so, argues Cauchy. Each domain—religion, the physical world, history, or mathematics—is fundamentally different and has its own sources and standards of truth. In particular, the truths of mathematics reside exclusively in the realm of mathematics and are derived solely from mathematical reasoning, paying no heed to physical considerations. In this realm mathematical reasoning is pure and rigorous, and the vagueness and uncertainty that had plagued the physical mathematics of the Enlightenment completely disappear.

True to his word, Cauchy then proceeds to construct a new kind of mathematics, strictly circumscribed, but pure and rigorous on its own terms. His goal is not to reach new results of the kind that his eighteenth-century predecessors had excelled at, but to construct a new architecture for mathematics, and for analysis in particular. The well-known results of the previous two centuries will now become building blocks of a new edifice: clearly bounded, requiring nothing from its physical surroundings, and logically rigorous in its internal relations. It would be an unassailable mathematical fortress, completely self-sufficient and impervious to the attacks of skeptics.[3]

The cornerstone of Cauchy's mathematical edifice is his definition of the basic concept of the *function*.[4] In the eighteenth century a function was understood to be shorthand for an "analytic expression," or simply an algebraic formula describing the relationship between different variables.[5] For Euler, d'Alembert, and Lagrange, this formulation seemed sufficient because they assumed that algebraic expressions were abstractions from a preexisting physical world, and such "functions" were therefore expressions of real relations.[6] For Cauchy, however, who seeks to separate mathematics from any physical underpinnings, this definition is meaningless; analytic expressions in themselves refer to nothing at all and are empty signs on a piece of paper. Cauchy therefore adopts a definition that relies not at all on algebraic expressions but instead focuses on simple universal relations between mathematical variables. After defining a "variable" as "a quantity that one considers as being able to receive successively several different values," he continues:

> If variable quantities are so joined between themselves that, the value of one of these being given, one can conclude the values of all the others, one ordinarily conceives these diverse quantities expressed by means of the one among them, which then takes the name of independent variable; and the other quantities expressed by means of the independent variable are those which one calls functions of this variable.[7]

For Cauchy, in other words, a function is simply the dependence of one variable on another. It has nothing to do with the complex notations of eighteenth-century algebra and is certainly not dependent in any way on the background of the physical world. All Cauchy needs for his function is a pristine world in which two types of objects exist: an independent variable and at least one dependent variable. Nothing else matters, and for all intents and purposes nothing else exists in this alternative world. The building of the mathematical universe has begun.

Making use of the simple concept of the variable, Cauchy then moves on to define an element that will be crucial for his reformulation of analysis: the concept of the limit. According to Cauchy, "When the values successively attributed to the same variable approach a fixed value indefinitely, in such a way as to end up by differing from it as little as one could wish, this last value is called the *limit* of all the others."[8] The limit, in other words, is independent of the series that leads up to it; it is not the "last term" of a series as it reaches infinity, as Newton argued and many eighteenth-century mathematicians assumed, at least implicitly. Rather, it is simply a value that the series can approach as closely as one wishes, even if the series never actually attains that value.

The significance of this definition does not lie in its novelty. Back in 1765 d'Alembert gave a very similar definition in the *Encyclopédie,* under the entry "Limite."[9] But d'Alembert, unlike Cauchy, was never satisfied with the notion that a series never reaches its limit, for d'Alembert believed that algebraic relations were expressions of actual physical relations in the material world. If a series never actually reaches its limit, it means that it never actually expresses a physical relation but only approximates it. As long as there is a gap, however small, between the series and its limit, one can argue that the calculus does not provide true results but only approximations, and that its results always contain an unbridgeable error. D'Alembert and his contemporaries assumed in practice that a series does reach its limit, and they based their calculations on this assumption, but they could never adequately respond to the charges of critics that the calculus contains a fundamental error. Bishop Berkeley's charge that infinitesimals were "ghosts of departed quantities" continued to haunt the grands géomètres of the eighteenth century.[10]

For Cauchy, however, the problem disappears altogether. Unlike his predecessors, for whom mathematical constructs necessarily corresponded to physical realities, Cauchy is interested only in the internal strength and coherence of his mathematical system. As a result, where d'Alembert and Euler, in order to preserve the correspondence of mathematics to the world, insisted that a series actually reaches its limit, Cauchy feels no pressure to follow their example. A limit, for Cauchy, is simply a number that a series approaches as closely as we choose—nothing more and nothing less. This may not be a satisfying definition for those who consider mathematics an abstraction from the physical world, but it is a perfectly clear and rigorous definition for those who, like Cauchy, consider mathematics to be a self-contained realm of rigor and truth. Cauchy then proceeds to deduce the entire edifice of eighteenth-century analysis and its main results from this simple definition of the limit. In a system such as Cauchy's, Bishop Berkeley's charge that there is always a difference between any term of an infinite series and its limit completely loses its force. "Of course there is a difference," we can imagine Cauchy responding; "I never claimed otherwise."[11]

The difference between Cauchy's approach and that of his predecessors is striking. The "grands géomètres" of the eighteenth century were certainly aware of the logical difficulties their position entailed, and some, most notably Lagrange, struggled with its implications. But they would never consider a solution like Cauchy's, which separated mathematics from its worldly underpinnings. That would have gone against their conviction that mathematics was an abstraction of the world, and would have delivered them into the hands of critics like Diderot, who argued that mathematics was so far removed from reality as to be irrelevant. For d'Alembert and his contemporaries, Cauchy's approach would have drained mathematics of all its content, making it, in essence, an empty game. It was, for them, a solution that was far worse than the problem of logical consistency.

It is clear from this that Cauchy's system was not a necessary development of his predecessors' approach, brought about by new mathematical discoveries. After all, d'Alembert and his contemporaries fully understood the definitions on which Cauchy would base his system and even stated them clearly. But unlike Cauchy, they chose not to base their mathematical approach or methods of calculation on these definitions, because doing so would run counter to their basic understanding of what mathematics is. The deep divide between Cauchy and d'Alembert is not one of mathematical innovation, in which Cauchy discovered new mathematical truths that were unknown to his predecessor. Rather, it is one of conception and of the deepest understanding of the nature of the field. D'Alembert believed that mathematics was an abstraction from the world, and that its relations described the deep interconnections underlying physical reality. Cauchy believed that mathematics is a self-contained realm that should be protected from external, nonmathematical interference. And therein lay the difference.

Cauchy now needs one more definition before moving on to what he considers the central concept of the calculus—the derivative. An "infinitesimal," or an "infinitely small quantity," is a variable that "decrease[s] indefinitely in such a way as to fall below any given number . . . A variable of this kind has zero for its limit."[12] The significant part of this definition is that for Cauchy an infinitesimal is not a magnitude in itself, but rather a variable that tends to zero. This is important because the founders of the calculus, and Leibniz in particular, clearly thought of infinitesimals as positive magnitudes, albeit the smallest ones possible. Although their successors in the eighteenth century offered different definitions of the term, they also assumed that infinitesimals had a real existence in the world. After all, if mathematics was the mirror of the material world, as

Enlightenment geometers believed, then certainly the central concept of the infinitesimal must be manifested in reality.

Cauchy's view, however, was different. He did not believe that mathematics was necessarily manifested in the world, and much of his work is, in fact, a sustained effort to keep the two realms clearly separated. The infinitesimal was for him a strictly mathematical notion that required no counterpart in the physical world and was freed from the intuitive notion of an "infinitely small" particle. It existed only in the mathematical realm and was defined solely in terms of the few concepts that belonged to that realm—variables, constants, and limits. As such, an infinitesimal for Cauchy was simply the name given to a variable that approaches as close to zero as one wishes. It has no actual magnitude in itself.

With this concept in hand, Cauchy moves on to define the derivative, which is the central concept of his reformulated calculus. By founding his system on the derivative, Cauchy is breaking with his predecessors' notion of the calculus as the art of calculating with "differentials." As understood by eighteenth-century practitioners, differentials were infinitely small magnitudes, and the goal of analysis was to determine their ratios and the ways they related to each other. Cauchy had little use for these logically suspect and quasi-material magnitudes and turned instead to the notion of the *derivative*—a term used by Lagrange before him. Unlike the differential, which is a static magnitude, a derivative is itself a function—a relationship between variables—related to the original function but not identical with it. As such, it is far better suited to Cauchy's notion of the calculus as a self-contained system of mathematical relationships than were the differentials, with their strong material connotations. Old habits die hard, however, and the eighteenth-century term *differential calculus* is in common use to this day, even as the field itself owes its substance to Cauchy's very different ideas.

Cauchy's definition of the derivative in *Cours d'analyse* is wordier than its modern equivalent, which relies almost exclusively on mathematical notation. But in essence, Cauchy's concept of the derivative is almost indistinguishable from the one taught to students of mathematics today:

> When the function $y = f(x)$ is continuous between two given limits of the variable x, and one assigns a value between these limits to the variable, an infinitely small increment of the variable produces an infinitely small increment in the function itself. Consequently, if we then set $\Delta x = i$, the two terms of the *difference quotient*

> $\frac{\Delta x}{\Delta y} = \frac{f(x+i)-f(x)}{i}$ will be infinitesimals. But while these terms tend to zero simultaneously, the ratio itself may converge to another limit, either positive or negative. This limit, when it exists, has a definite value for each particular value of x. But it varies with x . . . The form of the new function which serves as the limit of the ratio $(f(x + i) - f(x))/i$ will depend upon the form of the given formula $y = f(x)$. In order to indicate this dependence, we give to the new function the name derivative and we designate it . . . by the notation y' or $f'(x)$.[13]

The derivative, then, is a function that is defined at every point as the limit of the ratio between the change in the function and the change in the independent variable x as x moves toward its limit—zero. It is a dynamic definition, based entirely on the relations between changing variables rather than on ratios of static magnitudes like differentials. And thanks to Cauchy's careful preparatory work, his definition of the derivative is perfectly coherent and rigorous. The infinitesimals he refers to are not finite magnitudes but variables approaching zero, and the limit is not the last term of an infinite progression but a distinct value that a variable can approach as closely as one desires without ever reaching it.

Cauchy's definitions of the function, the limit, and the derivative are only preliminary steps in his project of reformulating the calculus as a self-contained mathematical system. With these in hand, he goes on to define all the basic components of the calculus and to prove all its known results. Higher-level derivatives, the mean-value theorem, integrals, and expansion series are all defined or proved and given their own place in the logical edifice that is Cauchy's calculus. Most of his basic concepts had been used before him or at least had some precedents in the eighteenth century.[14] But no one before Cauchy had attempted to make use of these concepts to create an insular mathematical world, accountable only to itself.

This was a radical project, completely at odds with the way mathematics had been practiced and understood by his predecessors. In the eighteenth century the notion that mathematics was its own distinct universe was reserved to critics like Buffon and Diderot, who argued that the field was too insular to be of use in understanding the world. In response, leading geometers such as d'Alembert and Euler took pains to demonstrate that mathematics was a science of the material world and could never be understood separately from it. To them, an approach that sought to separate mathematics systematically from the world confirmed, in effect, the charges of the enemies of mathematics. D'Alembert, Euler, and Lagrange may have possessed many of the separate ingredients that Cauchy used so effectively in building his system. But his project of reformulating analysis as a self-contained system of relations insulated from the world was not only foreign to them; it was abhorrent. Their successors, Fourier, Poisson, and the mathematicians of the École polytechnique, made sure that this Enlightenment tradition of mathematics remained in the ascendant in France throughout the nineteenth century. This, as we have seen, caused Cauchy no end of grief in his dealings with his colleagues at the École and the academy.[15]

The Mystery of the Quintic Equation

Meanwhile, as Cauchy was enduring social and professional ostracism at the hands of his unsympathetic colleagues, his young contemporary Évariste Galois was suffering much worse. Cauchy's troubles, as we have seen, were partly political, brought on by his advocacy of Catholic legitimism among his republican-minded colleagues. Galois' more severe crisis was also partly political but of an opposite kind, the result of his revolutionary republican activity. But inasmuch as their professional isolation was brought about not by politics or personality but by their mathematical views, Cauchy and Galois were undoubtedly on the same side. Both Cauchy and Galois were mathematical purists, advocating a new style of rigorous self-contained mathematics at a time when such views were rejected out of hand by the majority of professional mathematicians.

To understand the achievement of Galois, the ultimate tragic mathematician, we need to review the state of the theory of equations at this time. The story has its origins in the early sixteenth century, when flamboyant Italian virtuosi known as "Cossists" were competing with each other in the solution of numerical problems. Although the Cossists were pioneers in their time, they were already building on a substantial body of algebraic knowledge codified in the work of the third-century Alexandrian mathematician Diophantus and the Arab mathematicians who followed. In particular, the Cossists had inherited the universal solution for the quadratic equation, that is, in modern notation, $ax^2 + bx + c = 0$.[16]

As they well knew, and any high-school student knows today, the two solutions are given by the equation

$$x_{1,2} = \frac{-b \pm \sqrt{b^2 - 4ac}}{2a}.$$

It followed naturally that one of the Cossists' chief ambitions was to find a similarly universal solution for the cubic equation.

The first to make substantial strides in this direction was Scipione del Ferro (ca. 1456–1526), professor of mathematics in the University of Bologna, who developed a method for solving a certain class of the cubic.[17] Del Ferro did not publish his discovery but protected it as a great professional secret and only imparted it before his death to his student, the Venetian Antonio Maria Fiore. Meanwhile, another Cossist, Niccolò Tartaglia (1500–1557) of Brescia, was working on the problem of the cubic, and in 1530 he confided to a friend that he had discovered a method to solve certain types of the cubic. Hearing of this, but clearly believing his method superior, Fiore challenged Tartaglia to a public contest: on February 22, 1535, each was to deliver 30 solutions to the other's problems to a notary. The loser would stage 30 banquets for the winner.

In preparing for the contest, Tartaglia had meanwhile managed to solve all types of the cubic. He also had a good idea of which problems Fiore was capable of solving, and as a result he managed to solve all of Fiore's questions, while Fiore could solve none of the ones he posed. Clearly happy with his overwhelming victory, Tartaglia graciously waived the prize of 30 banquets, allowing his humiliated opponent to retire quietly from the scene. Like del Ferro before him, Tartaglia kept his methods secret, using them to raise his stature among the Cossists, and perhaps his income as well.

Tartaglia's victory over Fiore brought him to the attention of one of the great personages of the day, the Milanese Girolamo Cardano (1501–1576). A scholar, physician, gambler, and author of a celebrated autobiography, as well as a mathematician, Cardano at the time was working on a book on arithmetic. Intrigued by what he heard of Tartaglia's method, Cardano set about extracting the secret of the cubic from its reluctant author. In 1539, after many exchanges of letters and a visit to Cardano's home in Milan, Tartaglia finally yielded and entrusted the solution to Cardano in a mathematical poem. In order to preserve his priority rights, however, he made Cardano swear to keep the solution a secret and never publish it.

For the next few years Cardano kept true to his word, although he struggled mightily against its restrictions. Working with his assistant Ludovico Ferrari, he refined Tartaglia's method and also found a general solution for the quartic equation—an equation of the fourth degree. However, since all his work was based on Tartaglia's confidential disclosures, he could not publish. Then, in 1543, Cardano heard that it was not in fact Tartaglia but del Ferro who had been the first to crack the cubic. Cardano therefore traveled to Bologna, where he discussed the matter with del Ferro's successor at the university and examined the late master's papers. There he found evidence that del Ferro had cracked the secret several years before Tartaglia. This was the opening Cardano was looking for.

Since del Ferro had discovered the solution to the cubic before Tartaglia (although he had in truth discovered only part of it), Cardano reasoned that the secret of the cubic was not Tartaglia's to keep. The oath he had sworn was therefore null and void, and he was free to publish what he wished. When in 1545 Cardano published his book on algebra, the *Ars magna,* he therefore gave Tartaglia credit for independently finding the solution but insisted that the true and original discoverer was del Ferro. He then proceeded to provide the full solution of the cubic, as well as all the discoveries he and Ferrari had based on it.[18] Tartaglia was furious at this betrayal and even challenged Ferrari to a mathematical duel. But it was no use: the cat was out of the bag, the solution to the cubic was available to all, and Tartaglia had lost his precious advantage over his fellow Cossists.

What was the solution that Tartaglia was so desperate to protect, and Cardano eager to publish? It was, in essence, just like the known solution to the quadratic: a standard formula over an equation's coefficients that uses the extraction of roots to determine the solution to an equation. Such a formula is known as a *solution by radicals,* and it is the most powerful method for finding the roots of an equation: it is simple, easily memorized, and—most important—universally applicable. Just as any quadratic equation can be solved by its formula, so can the roots of any cubic equation be found through its solution by radicals.

In explaining this solution I will not use the language and notation of the Cossists, interesting and illuminating though they may be for the practices of early sixteenth-century mathematics. Instead, I will use modern notation, first developed by François Viète (1540–1603) and René Descartes (1596–1650) and standardized by the mid-eighteenth century into something not much different from what we use today. One reason for this is simply ease of understanding, since the Cossists' approach and terminology appear startlingly eccentric to one accustomed to modern notation.[19] The deeper reason is that my purpose here is not to study the Cossists and their approach to mathematics, but rather that of their successors nearly three centuries later. Lagrange, Abel, and Galois used essentially the same terminology and notation that we do today when referring to equations and the solution in radicals. When they discussed the work of Cardano and his peers, they did so in the terms presented here.

The first step in the resolution of the cubic is to show that any equation of the form $x^3 + Px^2 + Qx + R = 0$ has the same roots as a different equation that has no x^2 element and therefore takes the form $x^3 + px + q = 0$.[20] This is known as a *suppressed* cubic, and for any ordinary cubic there is a suppressed one that has the same roots. Once this translation is accomplished, one of the solutions to the equation—and the one that interested the Cossists most—is given by

$$x_1 = \sqrt[3]{\frac{-q + \sqrt{q^2 + (4p^3/27)}}{2}} + \sqrt[3]{\frac{-q - \sqrt{q^2 + (4p^3/27)}}{2}}.$$

This solution is always "real," meaning that it belongs to the domain of real numbers—integers, fractions, and their roots of any degree. But a cubic equation has three roots, not one, and the other two are more likely to be complex rather than real solutions. This means that they include the square roots of negative numbers and take the form $a + bi$, where a and b are real numbers and $i = \sqrt{-1}$. These solutions take the form

$$x_2 = \omega\sqrt[3]{\frac{-q + \sqrt{q^2 + (4p^3/27)}}{2}} + \omega^2\sqrt[3]{\frac{-q - \sqrt{q^2 + (4p^3/27)}}{2}} \text{ and}$$

$$x_3 = \omega^2\sqrt[3]{\frac{-q + \sqrt{q^2 + (4p^3/27)}}{2}} + \omega\sqrt[3]{\frac{-q - \sqrt{q^2 + (4p^3/27)}}{2}},$$

where $\omega = \frac{-1 + \sqrt{-3}}{2}$ and its square $\omega^2 = \frac{-1 + \sqrt{-3}}{2}$. The terms ω and ω^2 are known as complex roots of unity because they are solutions to the equation $x^3 - 1 = 0$. This is easily seen because $x^3 - 1 = (x - 1)(x^2 + x + 1)$, and 1 is clearly a root of unity because it is a root of $(x - 1)$, whereas ω and ω^2 are roots of $x^2 + x + 1$, meaning that $\omega^2 + \omega + 1 = 0$.

Cardano and his peers had no use for solutions that included such terms as $\sqrt{-3}$, which they considered meaningless. In general, they limited themselves to the real solution x_1 and discarded the problematic ones x_2 and x_3. An exception was Cardano's younger contemporary, the Bolognese Rafael Bombelli (1526–1572), who justified and used the roots of negative numbers in his *Algebra*. But Bombelli was the exception, and even through the seventeenth century most mathematicians remained wary of dabbling in such paradoxical expressions. When they were addressed at all, they were most often referred to derisively as "surd numbers." As it happens, however, it was precisely these surds that held the key to the next major development in the theory of equations.

In the years after Cardano's and Ferrari's solution of the quartic, most mathematicians believed that a general solution of the quintic (an equation of the fifth degree) was just around the corner. There seemed to be no reason why the quadratic, cubic, and quartic equations could be solved by radicals, whereas the quintic and equations of higher degrees could not. But as the decades wore on and no general solution was found, it became clear that the problem was far more difficult than initially imagined. The methods that had served del Ferro and Tartaglia well in solving the cubic failed to make a dent on the quintic, and the search was on for alternatives. Over the next two centuries numerous mathematicians tried their hand at this, producing valuable insights into the deep structure of equations and offering several new methods for solving the cubic and the quartic. Others produced practical numerical methods for solving equations of degree five or higher, using successive approximations that eventually lead to a solution of any desired accuracy. But a general solution by radicals remained seemingly just out of reach.

So things remained until the 1770s, when two French mathematicians turned their attention to the problem. One was a relative unknown, Alexandre-Théophile Vandermonde (1735–1796), who was known more for his musical talents than his mathematical work; the other was the great Lagrange himself. Vandermonde read his work to the Paris Academy of Sciences in November 1770, but the paper was not published in the academy's journal until several years later. As a result, Lagrange, who was at the Berlin Academy, apparently knew nothing of his investigations when in 1771 he published his authoritative work on the state of the theory of equations, "Réflexions sur la résolution algébrique des équations" (Reflections on the Algebraic Resolution of Equations).[21] Although they were produced independently of each other, the papers of Vandermonde and Lagrange share certain fundamental insights about the nature of equations and their solutions. But because Lagrange's work is both broader and more general, and because it was far better known to the mathematicians of his generation and those that followed, I here adhere to Lagrange's rather than to Vandermonde's presentation of the problem.[22]

"With regard to the solution of literal equations," Lagrange writes in the introduction to "Réflexions," "they have but little advanced from what they had been in the time of Cardano, the first to publish the solutions to the equations of the third and fourth degree." He then continues:

> The early success of the Italian analysts in this matter appeared to have been the last that could be made; it is no less certain that all the attempts that have been made to the present to advance the frontiers of this part of algebra, only served to find new methods for equations of the third and fourth degrees, none of which appear applicable, in general, to equations of a higher degree.[23]

The purpose of his treatise, he explains, is to examine the reasons for this persistent failure:

> I propose in this memoir to examine the different methods that have been found up to the present for the algebraic resolution of equations, of reducing them to general principles, and to clarify a priori why these methods are useful for the third and fourth degrees, and fail for higher degrees.[24]

In doing so, he concludes, he will not only "cast additional light on the solutions of the third and fourth degrees" but also serve as a guide to further attempts on the quintic "in furnishing [future mathematicians] with different views of this object, and sparing them a large number of futile steps and attempts."[25]

Lagrange begins his study of what makes an equation solvable by radicals by examining the familiar suppressed cubic equation $x^3 + px + q = 0$. Any cubic, as I have noted, can be reduced to this form. Suppose now that this equation has three different roots, designated x, y, and z, and we are looking at t, defined as $t = x + \omega y + \omega^2 z$, where ω and ω^2 are the cube roots of unity.[26] The value of t, Lagrange observes, depends on the order in which we insert the three roots into the algebraic expression $x + \omega y + \omega^2 z$, and since there are six different ways of arranging the roots, t acquires six different values. We can write the different values of t as follows:

$$\begin{aligned} t_1 &= x + \omega y + \omega^2 z, \\ t_2 &= z + \omega x + \omega^2 y, \\ t_3 &= y + \omega z + \omega^2 x, \\ t_4 &= x + \omega z + \omega^2 y, \\ t_5 &= y + \omega x + \omega^2 z, \text{ and} \\ t_6 &= z + \omega y + \omega^2 x. \end{aligned}$$

These six values of t, Lagrange continues, are the solutions to a sixth-degree equation, $f(X) = (X - t_1)(X - t_2)(X - t_3)(X - t_4)(X - t_5)(X - t_6) = 0$. Lagrange calls this equation a *resolvent,* and it is known as a Lagrange resolvent to this day.

Up to this point Lagrange does not seem to have made much headway in his efforts to solve the cubic. Starting with an equation of the third degree, he has now arrived at an equation of the sixth degree, a development that, despite the optimistic-sounding title "resolvent," can hardly count as progress. But $f(X)$, it turns out, is a very special function because the six values of t are not random but are related through ω, the cube root of unity. Let us take, for example, t_1 and multiply it by ω: $\omega t_1 = \omega(x + \omega y + \omega^2 z) = \omega x + \omega^2 y + \omega^3 z = z + \omega x + \omega^2 y = t_2$ (remember that ω is the cube root of unity, and therefore $\omega^3 = 1$). By repeating the process we find that the six different values of t can in fact be reduced to only two, t_1 and t_4:

$$t_2 = \omega t_1;\ t_3 = \omega^2 t_1;\ t_5 = \omega t_4;\ t_6 = \omega^2 t_4.$$

The resolvent can therefore be written as

$$f(X) = (X - t_1)(X - \omega t_1)(X - \omega^2 t_1)(X - t_4)(X - \omega t_4)(X - \omega^2 t_4) = 0.$$

If we now expand $f(X)$ by multiplying the factors $(X - t_k)$ with each other, and taking into account that $\omega^2 + \omega + 1 = 0$, we arrive at the resolvent equation

$$f(X) = (X^3 - t_1^3)(X^3 - t_4^3) = 0.$$ [27]

This is no longer a sixth-degree equation for X, but in fact a quadratic equation for X^3. If we expand the equation and denote X^3 by Y, we get

$$f(Y) = Y^2 - (t_1^3 + t_4^3)Y + t_1^3 t_4^3 = 0.$$

This is far more promising: beginning with an equation of the third degree, Lagrange has now reduced it to a far more manageable resolvent of the second degree, that is, a quadratic equation, in which *f(Y)* could now be solved if only we knew the values $(t_1^3 + t_4^3)$ and $t_1^3 t_4^3$. As it happens, however, their values are known and can be expressed in terms of p and q, the coefficients of the original equation. This is because $t_1^3 + t_4^3$ and $t_1^3 t_4^3$ are symmetric polynomials in the roots x, y, and z of the original equation $x^3 + px + q = 0$.[28] As Newton had shown, and as Lagrange and his contemporaries knew, symmetric polynomials in the roots of an equation can always be expressed as polynomials of the equation's coefficients. In this case, a certain amount of algebraic calculation shows that

$$t_1^3 + t_4^3 = -27q \text{ and } t_1^3 t_4^3 = -27p^3,$$

and *f(Y)* is therefore defined as

$$f(Y) = Y^2 + 27qY - 27p^3 = 0,$$

a simple quadratic equation that produces two values for Y and from them six values for X. These are the six roots of Lagrange's resolvent, which are the values of t_1 through t_6. Because these six ts are linear combinations of permutations of the roots x, y, and z of the original equation $x^3 + px + q = 0$, knowing the values of the ts easily leads to the values of the roots themselves. The cubic equation is thereby solved.[29]

All this might seem a very roundabout way to arrive at a solution that had been known for more than 200 years. But Lagrange was not looking for the solution to the cubic—he knew it, of course, well enough—but for a general algebraic method of solution. The solutions to the quadratic, cubic, and quartic equations had been known since the time of Cardano, but each seemed to be a unique case, relying on certain exceptional features of each type of equation. By basing his solution on the permutations of the roots of the equation multiplied by the roots of unity, Lagrange believed that he had found something different: a general method of solution that could be applied to equations of different degrees. If the general method leads to a solution, as it does in the case of the cubic, then the equation of that degree is solvable by radicals. But if the method fails and does not lead to a solution, then it might well be that no solution by radicals is possible.

Lagrange tried his method on equations of different degrees. The quadratic was easy because the only roots of unity are 1 and –1, and the resolvent leads almost immediately to the familiar formula for solving the quadratic.[30] The resolvent of the quartic seemed formidable at first, being an equation of degree 24, but it could be reduced to a bicubic equation and thereby solved.[31] But when Lagrange tried his method on the quintic, the results were different. Just as there are three cube roots of unity, so there are five quintic roots of unity. This means that there are 120 ways to arrange the five roots of the equation multiplied by the five roots of unity, which in turn leads to a resolvent equation of degree 120. Things do not end there, because in much the same way in which the resolvent of the cubic can be reduced from a sixth-degree equation to a quadratic, a quintic resolvent can be reduced from an equation of the 120th degree to one of the 24th degree. But that is where the process ends: Lagrange could see no way to reduce the resolvent further, much less to solve it. And so the general method that worked admirably for the quadratic, cubic, and quartic equations failed for the quintic.

Lagrange was not sure what to make of the failure of his method to resolve the quintic. In the introduction to "Réflexions" he wrote cautiously that he hoped that his work would help guide further attempts on the quintic and steer future investigators away from failed approaches. But Lagrange was clearly not optimistic about the prospects of finding a solution by radicals of the quintic. "Réflexions" represents his attempt to provide a general algebraic theory of equations, just as *Mécanique analytique* was his general algebraic theory of mechanics. In each instance Lagrange sought the deep mathematical principles from which all the particular laws of the field could be derived. In the case of the theory of equations he believed that he had discovered these underlying principles in the different permutations of the roots of the equation multiplied by the roots of unity, and in the resolvent that followed from them. And yet, he found, these deep and elegant principles failed to produce the desired solution for the quintic. What was one to think? Perhaps no solution was possible at all.

Although Lagrange was probably of the opinion that the quintic could not be solved by radicals, he never actually tried to prove this result. The reason was that Lagrange, as we have noted, believed that in the last analysis mathematical relations reflected the deep structure of the physical world. His role as he saw it was to reveal the profound mathematical harmonies that govern the world and to express them in the most elegant and general mathematical form. Lagrange was therefore very interested in describing the world and its relations as they actually existed. But proving that some hypothetical construct such as the general solution to the quintic did not exist? This, to Lagrange, was a waste of time. Since it did not describe anything in the world or add anything to our knowledge of its deep relations, proving such a theorem was mere sophistry. As a result, Lagrange, like other Enlightenment geometers, showed no interest in proving the existence or nonexistence of mathematical objects and relations.

But half a century after Lagrange's publication of "Réflexions," Abel took a very different view. In 1821, while still a student at the Christiania Cathedral School, Abel became convinced that he had discovered a solution to the quintic by radicals. He showed his work to his mentors, Holmboe and Hansteen, who then forwarded it to the mathematician Ferdinand Degen in Copenhagen. Degen was impressed, but when he requested some clarifications and examples, Abel came to realize that his solution was not as general as he had first thought. He therefore turned around and sought instead to prove that a solution to the quintic by radicals was simply not possible. By the end of 1823 he had succeeded, showing that no equation of a degree greater than four was solvable by radicals. He quickly printed his solution in abbreviated form, hoping that this spectacular result would serve as an introduction to the great mathematicians of Europe in his upcoming tour of the Continent.[32] In this, as we have seen, he was largely disappointed.

The contrast between the attitudes of Lagrange and Abel here is revealing. Lagrange, once he had failed to resolve the quintic, was at the end of the road and turned his attention to other fields. Abel, when he realized the errors in his early "solution" of the quintic, turned around and tried to prove that no solution was in fact possible. This points not only to a personal difference in the interests of Lagrange and Abel, but also to a generational divide: Lagrange, like his Enlightenment contemporaries, was interested in mathematical relations to the extent that they reflected an underlying worldly reality. But Abel, like Cauchy, Galois, and most modern mathematicians, was interested in mathematical relations for their own sake. To Abel, showing that a solution did not and could not exist was just as significant as discovering a solution. Both results provided invaluable information about the contours of mathematical reality, a reality independent of the world around us and governed only by its own mathematical rules.

The final chapter in the centuries-long story of the quintic was written by none other than Galois, the tragic genius who saw the eternal cypress close in on him in his last days. Galois worked independently of Abel and learned of the Norwegian's accomplishment only after Abel was posthumously awarded the Paris Academy's grand prize in 1830. Nevertheless, just as the legend of the life and death of young Galois echoes the hagiographic tales told of Abel after his death, so did Galois' approach to the problem of the quintic resemble that of Abel. From Cardano to Lagrange, mathematicians had been searching for the general solution to the quintic equation. Abel had shifted the focus of the question by looking not for an actual solution but for the structural characteristics of an equation that make a solution by radicals possible—or impossible. Galois, like Abel, was interested in the hidden architecture of equations rather than in their solution, but he went much further than his Norwegian contemporary. Abel had shown that for deep structural reasons the quintic cannot be solved by radicals; Galois revealed the deep structures that determine whether any equation of any degree is solvable by radicals.[33]

A brief clarification is in order. Mathematicians from del Ferro to Abel were looking for general solutions by radicals to general equations of the form $a_1x^n + a_2x^{n-1} + a_3x^{n-2} + \ldots + a_nx + a_{n+1} = 0$, where the coefficients a_k are rational numbers (integers and fractions). Cardano and Ferrari had found the general solution for all equations for which $n \leq 4$, and Abel had shown that no general solution is possible when $n \geq 5$. This, however, does not mean that specific equations of degree 5 or greater are not solvable by radicals. For example, although the general equation of degree 6 is not solvable by radicals, the simple equation $x^6 - 1 = 0$ clearly is. The solution is $x = \sqrt[6]{1}$, which gives the six values of the sixth roots of unity. Similarly, certain equations of any degree are solvable by radicals, even if the general equation of that degree is not. What Galois found was the deep architecture that makes any particular equation—not just the general one—solvable by radicals or not. In principle one could use Galois' method to determine whether a particular equation of degree 37 is solvable, although in practice the calculations would be unmanageable. Impracticality, however, concerned Galois not at all: he was interested in the deep structure of equations, not in the fortunes of any particular one.[34]

When Galois asked about the solvability of an equation, he did not ask simply whether an equation could be solved or not, but rather what kind of a solution it might have. For example, the equation $x^2 - 2 = 0$ has rational coefficients (1 and -2) but no rational solution. The equation's solutions, $\sqrt{2}$ and $-\sqrt{2}$, are irrational and belong to the field of real numbers, which is larger than the field of rational numbers.[35] As it happens, however, we do not need to move to the entire field of real numbers in order to find the solution to our equation. In fact, Galois observed, it is sufficient to add a single element, $\sqrt{2}$, to the existing field of rational numbers in order to obtain the field of solutions. If we call the field of rational numbers by its modern designation Q, then Q($\sqrt{2}$) (that is, Q with the addition of $\sqrt{2}$) is a field as well, and it contains all the roots of the equation $x^2 - 2 = 0$.

Galois' insight was that this is true not just for a simple equation such as $x^2 - 2 = 0$, but for any equation of any degree that takes the form $a_1x^n + a_2x^{n-1} + \ldots + a_nx + a_{n+1} = 0$. If the coefficients a_k are in a field designated K, then there exists a field L larger than K that is the smallest field that contains all the roots of the equation. If the equation is solvable by radicals, this means that the solution contains several nested roots, and L can be obtained in the following way: first, the solution calls for the extraction of a root of, say, the pth degree of a value in K. If this pth root is not already in K (as it rarely is), then it is adjoined to K. Let us call the resulting field, which is larger than K, K'.

The next stage in the solution calls once again for the extraction of a root of p_1th degree from a value in K'. By adding this p_1th root to K' we get a new and larger field, K'', and so on. So we move step by step through all the nested roots of the solution by radicals, adding the roots one by one to the K fields. By the time we complete the operations, we have created the field $K^{(\mu)}$, which contains all the values encountered in the solution by radicals, including, ultimately, the solutions themselves. The key to finding whether a given equation is solvable by radicals, Galois reasoned, is to see whether the field of solutions L can indeed be created in this manner. If the answer is yes, then the equation is indeed solvable by radicals, and $L = K^{(\mu)}$; but if the answer is no, then no such solution is possible.

All this sounds reasonable enough and is, in fact, provable, but some major challenges still remain. First, how does one characterize this field L that contains all the roots of the original equation? Second, once the field is characterized, how does one determine whether it can be formed through the repeated addition of radicals to the field of coefficients K?

One of Galois' chief achievements was formulating answers to both these questions and proving his results.[36]

To characterize the field, Galois proceeded in a manner reminiscent of Lagrange but more general. Given an equation of the *n*th degree with *n* different solutions, Galois looked at the expression $T = Aa + Bb + Cc + \ldots$, where *a, b, c*, and so on are all the roots of the equation and *A, B, C*, and so on are integer coefficients.[37] Now if we change the order of the roots in this expression (say, switch *a* with *c*, *b* with *d*, and so on), then the value of *T* changes as well. Because there are *n!* different ways to arrange the roots, then *T* can take on a maximum of *n!* different values.[38] Galois showed that it is always possible to select the coefficients *A, B, C*, and so on in such a way that *T* takes on *n!* different values, $t_1, t_2, t_3, \ldots, t_{n!}$. Because *a, b, c*, and so on are roots of the original equation, all the values of *t* belong to the field of the solutions *L*. In fact, it can be shown that adding any one of the values of *t* to the field of coefficients *K* is sufficient to produce the entire field of solutions. In mathematical notation this means that $L = K(a,b,c, \ldots) = K(t)$.

In order to characterize this field of solutions, Galois then moved to create a resolvent equation in the manner of Lagrange. Given the *n!* different values of *t*, he posited an equation

$$F(X) = (X - t_1)(X - t_2)(X - t_3) \ldots (X - t_{n!})$$

whose roots are the *n!* different values of *t*. Once expanded, *F(X)* turns out to be a polynomial of degree *n!*,

$$F(X) = a_1X^{n!} + a_2X^{n!-1} + \cdots + a_{n!}X + a_{n!+1}.$$

Notably, however, the polynomial *F(X)* has this unique quality: its coefficients all belong to *K*, the field of the coefficients of the original equation. This is despite the fact that $a_1, \ldots, a_{n!+1}$ are all products of $t_1, \ldots, t_{n!}$, which belong to the larger field *K(t)* that contains the solutions to the original equation. The reason for this is that $a_1, \ldots, a_{n!+1}$ are systematic products of $t_1, \ldots, t_{n!}$, which themselves are linear combinations of the *n!* possible permutations of the roots *a, b, c*, . . . of the original equation. As such, the coefficients $a_1, \ldots, a_{n!+1}$ are symmetric polynomials of the roots of the original equations, and as I have already noted in discussing Lagrange's work, symmetric polynomials of the solutions can always be expressed as polynomials in the original function's coefficients. And because these belong to *K*, so do $a_1, \ldots, a_{n!+1}$, the coefficients of *F(X)*.

Up to this point the parallels between Galois' approach and that of Lagrange before him are clear. Lagrange also looked at the linear combinations of the roots of the original equation and at their changing values t when their order was changed in all possible ways (although, unlike Galois, he insisted that the coefficients of the different combinations be roots of unity). He then created a resolvent equation for the cubic $F(X) = (X - t_1)(X - t_2) \ldots (X - t_6) = 0$ that he elegantly expanded, making use of the unique characteristics of the roots of unity. He then showed that the coefficients of the resulting equation, $(t_1^3 + t_4^3)$ and $t_1^3 t_4^3$, are symmetric polynomials in the roots of the original equation and can therefore be expressed as polynomials of the coefficients of the original equation.

At this point, however, the paths of Lagrange and Galois diverged: Lagrange wanted to solve the resolvent equation, which he was able to do for the cubic and the quartic, but not for the quintic. Galois, in contrast, was not interested in a solution for the resolvent but instead wanted to use it in order to characterize its field *K(t)*, which is identical with *L*, the field of solutions of the original equation. In essence, whereas Lagrange was looking for a general solution to equations of different degrees, Galois was studying the inner structure that made such solutions possible.

Starting with the resolvent $F(X) = a_1 X^{n!} + a_2 X^{n!-1} + \ldots + a_{n!} X + a_{n!+1}$ (whose coefficients a_k are in K), Galois decomposed it into its factors $F(X) = G_1(X)\ G_2(X) \ldots G_s(X)$, so that each $G_i(X)$ is irreducible and has a coefficient in K—meaning that it cannot be factored further over K. At the same time, because each $G_i(X)$ is a factor of $F(X)$ (whose roots are the $n!$ values of t), each value of t is also a root of one of the $G_i(X)$s. Each $G_i(X)$ is therefore an irreducible equation with coefficients in K, each of whose roots produce *K(t)*—the field of solutions of the original equation.

Next, Galois associated the resolvent $F(X)$ with a group of permutations of the solutions $a, b, c, \ldots$ of the original equation in the following manner. Taking any t_i at random, he noted that it is a root of one of the $G_i(X)$s (because $G(X)$ is a factor of $F(X)$), and that this $G_i(X)$ is of degree r and can be shown to have r distinct roots. Now because t_i belongs to the field of solutions of the original equation, each of the n solutions $a_1, a_2, a_3, \ldots, a_n$ can be expressed as a rational function of t_i. In this manner $a_1 = h_1(t_i)$, $a_2 = h_2(t_i)$, and $a_n = h_n(t_i)$. But what happens if t_i is replaced in these functions by any of the $r - 1$ other values? In that case, it turns out, the functions once again produce all the n values of a, but in a different order. And so, by running through all r values of t that are roots of $G(X)$, Galois produced the following group of permutations:

$$h_1(t_1), h_2(t_1), h_3(t_1), \ldots, h_n(t_1),$$
$$h_1(t_2), h_2(t_2), h_3(t_2), \ldots, h_n(t_2),$$
$$\ldots$$
$$h_1(t_r), h_2(t_r), h_3(t_r), \ldots, h_n(t_r).$$

Each element $h_i(t_k)$ is one of the solutions a_i of the original equation, and in each row they appear in a different order. The number of rows r corresponds to the number of roots of $G(X)$. This group of permutations of the roots is known today as the *Galois group* of an equation, and it can be shown that it comes out the same regardless of the choice of $G(X)$ and its root t_i.

What happens to the Galois group if we enlarge K by adding a new element to the field? In that case $G(X)$, which was irreducible over K, may in fact be reducible over the larger field K' into polynomials of a lower degree, $G(X) = g(X)h(X)$. This means that the root t_i of $G(X)$ is now a root of an equation of a lower degree r', which has only r' roots. The new Galois group will therefore be a subgroup of the original one, with only r' rows instead of r rows. Therefore, as K expands, the Galois group either remains the same (if $G(X)$ is still irreducible) or grows smaller. When K finally becomes L, the field of solutions, t_i becomes the root of an equation of the first degree—a linear equation. Since t_i is its only solution, the Galois group has only a single row. It is reduced to the identity substitution.

We have already noted that if an equation is solvable by radicals, then it is possible to move step by step from its field of coefficients K to its field of solutions L in a particular manner: if K can be expanded to K' by adjoining a radical in K, and K' can be expanded into K'' by adjoining a radical in K', and so on, step by step, until the field of solutions L is attained, then the equation is solvable by radicals. Galois showed that with each of these steps, as K is expanded to K', K'', and so on, the Galois group grows smaller or stays the same. Finally, when the field of solutions L is reached, the Galois group is reduced to the identity substitution alone. In other words, if an equation is solvable by radicals, then its Galois group can be reduced in this way.

But the reverse is also true: if a Galois group G of an equation can be reduced step by step to subgroups G', G'', and so on, with each $G^{(k)}$ being a normal subgroup of its predecessor $G^{(k-1)}$, until only the identity substitution is left, then the equation is solvable by radicals.[39] This is because each successive normal subgroup $G^{(i)}$ corresponds to a field $K^{(i)}$ and is the original equation's Galois group over that field. Finally, the identity substitution is the Galois group of the field of solutions L. When a Galois group can be reduced to the identity substitution step by step in this manner, then the group is called solvable. Accordingly, if the Galois group of an equation is solvable, then the equation itself can be solved by radicals.

To review, at the beginning of this discussion we were looking for a criterion that would tell us whether the field of solutions L can be obtained from K, the field of the coefficients of the original equation, by successive additions of radicals. In other words, if K can be expanded to K' by adjoining a radical to K, and K' can be expanded to K'' by adjoining a radical to K', and so on, step by step, until the field of solutions L is attained, then the equation is solvable by radicals. This, Galois showed, was precisely what the Galois group tested: the group is solvable if and only if the field of solutions L can be obtained from K by a step-by-step process of adjoining radicals. In other words, Galois' criterion is this: an equation is solvable by radicals if and only if its Galois group is solvable. Q.E.D.

An Impractical Solution

Such is Galois' answer to the question of how to tell whether an equation is solvable by radicals. To modern mathematicians it is a clear and definitive answer that looks to the very heart of the theory of equations and closes the book on a problem that had troubled mathematicians since the days of Cardano. "Galois ends the story" is how one recent book dealing with the search for the solution to the quintic equation characterizes Galois' achievement, and that is indeed how his work is seen today.[40] To his contemporaries, however, the question seemed far from settled, for Galois' solution appeared to suffer from a fatal flaw: it was completely useless.

No one, in fact, seemed more aware of this problem than young Galois himself. Already in 1830, at the age of 18, he wrote in an unpublished introduction to a projected publication of his works:

> If you now give me an equation that you have chosen at your pleasure, and that you desire to know whether it is resolvable by radicals or not, I could do nothing but indicate to you the means of responding to your question, without wishing to charge either myself or anyone with the task of doing it. In a word, the calculations are impractical.[41]

It follows that Galois' subtle method for determining whether an equation is solvable by radicals was, in fact, all but useless for performing that very task. "It appears from this that no fruit is derived from the solution that we propose," Galois continued rather sheepishly in the same preface. It is a startling admission from one who is hailed today as the founder of the modern field of algebra.

We should not therefore be surprised at the puzzlement and incomprehension of the mathematical grandees of the academy when they were confronted with Galois' work. The Italian Cossists, as they well knew, had solved the cubic and quartic equations by radicals three centuries before, but could go no further. Lagrange, some decades before, had also delved into the mysteries of the theory of equations: he had shown why cubic and quartic equations were amenable to a solution by radicals, whereas the case of the quintic was fundamentally different, far more difficult, and might not have such a solution at all. Finally, Abel, writing only a few years before Galois, had shown that indeed no solution by radicals was possible for the general quintic equation. Each of these, from Tartaglia to Abel, had revealed something concrete about the nature of equations and how to solve them.

But what had Galois done? He had not found a new solution in the manner of the Cossists, discovered a fundamental truth about equations of different degrees in the manner of Lagrange, or proved this truth, as Abel had done. What he had found was a method to determine whether any particular equation was solvable by radicals, and that method—by his own admission—was next to useless. What, then, was the point of the whole exercise? We should perhaps have some sympathy for Poisson, who, recognizing young Galois' brilliance, had reached out to him and asked that he write a memoir for the academy. But when he actually read through the young man's treatise, Poisson was left with more questions than answers: "We have made all possible efforts to understand M. Galois' proof," he wrote, adding that "his reasonings are neither sufficiently clear, nor sufficiently developed for us to give an opinion in this report."[42] The exasperated puzzlement evident in these comments did not arise from an inability to comprehend Galois' technical argument, which Poisson's detailed comments show he understood well enough. The misunderstanding went deeper: what, Poisson seemed to be asking, was Galois trying to do?

The beginning of Galois' answer can be found a little later in the 1830 introduction, where he explains his disdain for the elaborate calculations that would make his method "practical." "There exists in effect for these kinds of questions, a certain order of metaphysical considerations that soar above all calculations and often makes them useless," he wrote. It is these higher considerations, not the specific calculations, that Galois is after: "All that makes this theory beautiful, and in truth, difficult, is that one has always to indicate the course of analysis and foresee its result without ever being able to perform the calculations."[43] In other words, the fact that the theory is useless for calculating specific results is not a drawback for Galois, but rather an advantage. Like Jacobi, who some decades later proclaimed that "it is the glory of science to be of no use," Galois revels in the impracticality of his results, which keeps them pure and beautiful, free of the messy, grungy work of prolonged calculations.[44] And perhaps it is no wonder that the young man who styled himself as a romantic hero, who mused about the proximity of death and sacrificed his life for love and liberty, would always value beauty over practicality.

Galois provided a more detailed exposition of his views on mathematics in another unpublished work, the "Préface" he composed in December 1831 while a prisoner at Sainte-Pélagie. The advanced and elegant techniques developed by mathematicians from Euler to Lagrange, Galois argued, were aimed at simplifying complex problems by dealing with many operations at once. "Now I believe that the simplifications produced by the elegance of calculations . . . have their limits," he continued. "I believe that the moment will come when the algebraic transformations foreseen by the analysts will find neither the time nor the place to be executed; at that point it will become necessary to be content with having foreseen them."[45]

The avoidance of detailed calculations, and the fact that his method was practically useless for obtaining specific results, were not drawbacks in Galois' eyes. Rather, they pointed the way toward a new mathematics that would dispense with such pedantry. "Jump with both feet over the calculations," he advised his readers; "group the operations, classify them according to their difficulty, not according to their form. Such is, according to me, the mission of future geometers; such is the road that I entered upon in this work."[46] The future of mathematics, according to Galois, lay not in producing specific results and resolving specific problems in the manner of his Enlightenment predecessors. These investigations had assumed that the subject matter of mathematics is ultimately the material world, and the purpose of the field was to reveal its hidden harmonies. The new mathematics, he argued, would distance itself from such concerns and forge its own path. The mathematicians of the future, Galois believed, will be concerned only with the beautiful inner relations that govern the mathematical world itself, paying no heed to the untidy demands of physical reality or any particular results. The mathematics of the future will be an "analysis of analysis," and his own beautiful but impractical work was but a harbinger of what it would be like.[47]

Remarkably, for a dejected loner working on the margins of society, Galois was right. Over the following decades, as the center of mathematical learning slowly shifted from Paris to Göttingen and Berlin, the field acquired precisely the character that Galois had predicted. Abandoning mathematics' roots in the sciences and the physical world, academic mathematicians were no longer concerned with producing numerical results or answers to the specific questions that had occupied their predecessors. Euler and d'Alembert had been concerned with vibrating strings, and Fourier had attempted to describe mathematically the dissemination of heat, but their nineteenth-century successors had loftier goals. They devoted their efforts to investigating the contours and inner relations of the mathematical world itself, engaging in what the young Frenchman called an "analysis of analysis." Galois' vision of the future of mathematics had in fact come true.

Composing his "Préface" in prison, just a few months before his death, Galois saw himself not just as a victim of unjust persecution but also as a prophet of the future of mathematics. And so it proved: writing in 1962, the French mathematician Jean Dieudonné remarked on the "strangely modern allure" of Galois' thought. "His insistence on the conceptual character of mathematics, his aversion to long calculations that obscure direct ideas, his care to group the problems according to their profound structural affinities rather than their superficial aspects, all of this is now familiar to us."[48]

Dieudonné was a member of Bourbaki, a collective of French mathematicians whose stated goal was to reformulate all of mathematics as a self-contained and strict axiomatic system. As a result, it is hardly surprising that Dieudonné would find Galois' reasoning remarkably "modern" in comparison with that of his contemporaries. For Galois had indeed pointed the way toward the type of mathematics cherished by Bourbaki: highly abstract, focused only on inner mathematical relations, and unconcerned with specific numerical results.

Viewed in this light, Galois may well have been the most radical of a group of mathematicians who changed the face of the field in the early nineteenth century. Cauchy, Abel, Galois, and Jacobi shared the belief that the old mathematics had reached a dead end, and the field now required a new beginning. The new mathematics, they insisted, would no longer concern itself with specific numerical questions derived ultimately, even if remotely, from physical reality. Instead, mathematics would now be its own separate world, a realm of absolute truth governed solely by strict and rigorous reasoning. If the results ultimately corresponded to certain physical phenomena and could be useful in describing them, that was well and good. But it had no bearing on their truth, for the value of mathematical results would be judged by internal mathematical criteria alone. As Galois, Jacobi, and later Hardy all noted, some of the most beautiful and important mathematical creations are also the most useless. This does not detract from the theories' importance as mathematical creations; to the contrary, it adds luster to them as the beautiful embodiments of pure and unsurpassed truth.[49]

Nowhere is the transition from the "real" world to the mathematical one, from the useful and concrete to the useless and beautiful, more in evidence than in the emergence of non-Euclidean geometry in the early nineteenth century. It was a radical innovation that turned on its head the most cherished assumptions not only of mathematics but of the entire order of knowledge, challenging the very notion of truth. Euclidean geometry, long the model of undisputed knowledge and the envy of more contentious fields, was replaced by an infinite number of alternative geometries, mutually exclusive but all equally true. Geometry, the bedrock of certain knowledge was set adrift from our world and became a self-contained universe all its own—perfect, logical, even beautiful, but completely unrelated to the real world we live in.

Because of geometry's status as the bedrock of classical mathematics, it is its transformation that most completely captures the transition from the "natural" mathematics of the Enlightenment to the "otherworldly" mathematics of the early nineteenth century. By tracing the evolution—or rather revolution—in the understanding of geometry and its relationship to the world, we can also trace the transformation of mathematics as a whole in those years. Most important, once again the transition is accompanied by a tragic tale of unacknowledged genius that is strongly reminiscent of the legends of Galois and Abel. It is the story of a young romantic hero who made remarkable discoveries and suffered for them in the same years in which Galois and Abel lived and died. His name was János Bolyai.

IV

A NEW AND DIFFERENT WORLD

8

THE GIFTED SWORDSMAN

Euclid's Blemish

For more than two millennia the edifice of Euclidean geometry was hailed as the very model of a true science, the purest example of proper logical reasoning. Its rigorous deductive method, in which necessary conclusions followed inexorably from unerring assumptions, set a standard to which other fields could only aspire. From theology to natural philosophy, all fields of knowledge seemed trapped in endless controversies that could never be properly decided. Only in Euclidean geometry could questions be settled with absolute certainty through proper reasoning and clear demonstrations. "The theorems of Euclid . . . still today as for many years past, retain in the schools their true purity, and their strong and firm demonstrations," wrote Christopher Clavius, founder of the Jesuit mathematical tradition in the 1580s. For him, as for many other early modern scholars, Euclidean geometry was much more than the science of space. It was, in fact, the keystone to the entire order of knowledge.[1]

Nevertheless, it was acknowledged, even Euclid's masterpiece was not without blemish. From time to time, beginning with Euclid's own contemporaries, geometers and philosophers had challenged various aspects of his work and offered alternatives. This did not mean that they objected to the *Elements* as a whole or doubted its conclusions, but quite the opposite: fully accepting the truths and the methods of Euclid's masterpiece, they felt that certain isolated sections of the *Elements* did not live up to the lofty standards of the work as a whole and should therefore be improved upon. And nothing in Euclid's geometry attracted as much criticism and controversy as the fifth postulate, also known as the parallel postulate, which states the following:

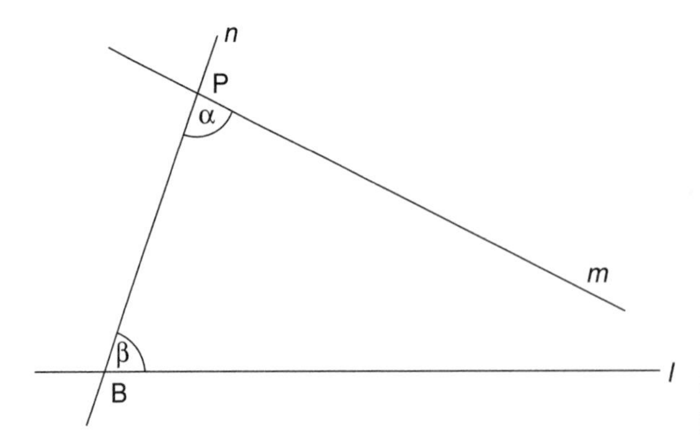

Figure 8.1. Euclid's parallel postulate.

> If a straight line falling on two straight lines make the interior angles on the same side less than two right angles, the two straight lines, if produced indefinitely, meet on the side on which are the angles less than two right angles.[2]

To understand what this means, look at Figure 8.1, where the straight line n falls on the two straight lines l and m, creating interior angles α and β. The postulate states that if the sum of the angles $\alpha + \beta$ is smaller than two right angles (or, as we would say, 180°), then the two lines l and m will meet on the same side of n as the interior angles (in this case, on the right side).

This seems intuitively true: we can "see" that if α and β are small enough, the two lines l and m will indeed meet on the right side. But this in itself does not tell us why Euclid chose to include this plausible hypothesis as one of his basic postulates. The reason is that the postulate implicitly defines parallel lines, and parallels are indispensable for nearly all the nontrivial results of Euclidean geometry.

Parallels, according to Euclid, are two straight lines in the plane that never meet, even if they are extended indefinitely. According to the fifth postulate, the lines l and m in Figure 8.1 cannot be parallel if $\alpha + \beta < 180°$, because then the lines meet on the right side of n. At the same time l and m cannot be parallel if $\alpha + \beta > 180°$ either, because then the interior angles on the left side of n will be smaller than 180°, and the lines l and m will meet on the left side of n. This leads to the inevitable conclusion that lines l and m are parallel if and only if $\alpha + \beta = 180°$.

That this was an extremely important result was as clear to Euclid as to his critics. Knowing the relationship between interior angles and parallel lines makes it possible to "move" equal angles around and to recreate them in different locations of a geometrical construction. For example, the proof that the sum of the angles of a triangle is always 180° relies on transporting two of the angles next to the third by constructing just such a parallel. When the three angles of the triangle are placed side by side in this manner, it becomes clear that their sum is two right angles. Nevertheless, the fact that the fifth postulate was true, and indeed very useful, did not convince Euclid's critics that it should be a postulate.

Postulates are the cornerstones of Euclid's geometry, on which the entire edifice stands or falls. In Book 1 of the *Elements* they come immediately after the definitions of objects, such as angles and polygons, and before simple rules of reasoning, such as "things which are equal to the same thing are also equal to one another."[3] The postulates state, for example, that a line can be drawn between any two points, and that any line can be extended. They are simple statements, self-evident and universally accepted, which no rational person would possibly deny. The rest of the *Elements* consists of drawing the necessary conclusions from the postulates, and as long as they are correct, all the theorems that follow are true. But if there is a flaw in the postulates, then all the layers of deduction that follow might also be unsound. In that case the brilliant temple of geometry—the model of correct reasoning and the standard of truth for all fields of knowledge—would be in danger of crumbling to the ground.

The stakes for selecting the correct postulates are therefore extremely high, and when viewed in this light, the parallel postulate is clearly in trouble. Any rational person would undoubtedly agree that a line can be drawn between two points. But would he or she also agree that "if a straight line falling on two straight lines makes the interior angles on the same side less than two right angles, the two straight lines, if produced indefinitely, meet on the side on which the angles are less than two right angles"? This is a far more complex statement than claiming that a line can be extended indefinitely, and it is not nearly as self-evident. The problem was not that the fifth postulate might be wrong, since practically all of Euclid's commentators up to the nineteenth century never doubted that it was true. Rather, it was qualitatively different from the other postulates and in fact did not look like a postulate at all. It looked much more like a theorem that must be proved on the basis of simpler preceding assumptions. Therefore, for two millennia geometers of different creeds and backgrounds tried repeatedly to "prove" the fifth postulate. If this ugly duckling among postulates could be deduced from the others, they reasoned, all of geometry would be placed on firmer ground.[4]

Most of what we know of the early attempts to prove the parallel postulate comes from the Greek philosopher and mathematician Proclus (410–485 CE), who among his many works authored a *Commentary on the First Book of Euclid.* Proclus presented the arguments of a whole series of ancient commentators who tried to show that the postulate could be derived from Euclid's more natural assumptions. These ranged from the relatively obscure, such as Posidonius and Geminus, who lived in the first century BCE, to the great Ptolemy of Alexandria, author of the *Almagest* and the *Geography,* who lived in the second century CE. Proclus pointed out the flaws in all their arguments, but this does not mean that he considered the attempts to prove the postulate futile. To the contrary, he firmly believed that the parallel postulate is, in fact, a theorem, and he proceeded to offer his own "proof." Needless to say, just like the arguments he critiqued, Proclus's own argument was found lacking.[5]

In the following centuries the controversy over the fifth postulate was kept alive through the Arabic translations of Euclid and his commentators. Medieval Persian scholars in particular, including the celebrated Omar Khayyam (1048–1122) and Nasir Eddin al-Tusi (1201–1274), suggested new ways of deducing the fifth postulate from the other postulates and definitions. None of these proofs was widely accepted, but the impact of the Persian work was substantial. When interest in the problem revived in Christian Europe in the seventeenth century, John Wallis (1616–1703), who was Savilian Professor of Mathematics at Oxford, commissioned a Latin translation of al-Tusi's work.[6]

Wallis was likely the first European geometer of the seventeenth century to engage with the problem, and in 1693 he published his own creative "proof" of the parallel postulate. Wallis's reasoning relied on the fact that similar geometrical figures exist, that is, that two figures can have the same angles and consequently the same shape but be of different sizes. This assumption seems unexceptionable to us since it corresponds to our experience of the world, as well as what we were taught in geometry. Furthermore, Wallis's reasoning was perfectly correct, and one can indeed prove the fifth postulate by assuming the existence of similar figures. But was anything truly gained by replacing the parallel postulate with the assumption of similar figures? Most geometers thought not. Both assumptions are perfectly true, they thought, and intuitively compelling, but both are qualitatively different and more complex than Euclid's other postulates. What Wallis had actually shown was that the parallel postulate implied the existence of similar figures, and that similar figures implied the parallel postulate. He could not, however, prove either of these claims from Euclid's more basic assumptions.

The Jesuit Geometer

Wallis did not worry too much about this deficiency in his demonstration. Famous even in his own lifetime for his intuitive and nonrigorous approach to mathematics, Wallis sought to deduce mathematical truths directly from worldly experience.[7] Since similar triangles are for all intents and purposes a fact of life, a proof based on their existence was, for Wallis, perfectly adequate. Things were different, however, for the Italian Girolamo Saccheri (1667–1733), the geometer who made the most systematic attempt in that era to prove the parallel postulate. Saccheri was professor of mathematics at the University of Pavia, but more important, he was a Jesuit and a loyal adherent of that order's mathematical tradition.

For Jesuit mathematicians from Clavius onward, Euclidean geometry was a model of true knowledge to which all other sciences should aspire. Far more than a subfield of mathematics, it presented a picture of how all knowledge should be pursued—moving irresistibly from simple assumptions to necessary and irrevocable conclusions. As such, geometry was a key component of the Jesuits' broader views on the nature of knowledge and, by implication, on the religious and social order as well. Any perceived weaknesses in the Euclidean system of logical deduction could potentially undermine the Jesuit worldview of a fixed and hierarchical world order. For Saccheri and his brethren, it was therefore imperative that doubts about Euclid be put to rest. He accordingly called his book *Euclides ab omni naevo vindicatus, sive conatus geometricus quo stabiliuntur prima ipsa universae geometriae principia* (Euclid Vindicated from All Blemish; or, A Geometrical Work in Which the First Principles of Universal Geometry Are Secured).[8]

Unlike his predecessors, who tried to prove the parallel postulate directly from Euclid's other premises, Saccheri approached the question in a roundabout manner. Let us assume, he argued, that the parallel postulate is false and draw necessary conclusions from this. If the postulate is, in fact, true (as Saccheri never doubted it was), then the false assumption will inevitably lead to a logical contradiction at some point down the road. From this we will necessarily conclude that our original assumption was wrong, and hence the parallel postulate is true. Saccheri adopted this method in the hope that the problem that had proved impervious to a direct assault would finally succumb to an indirect one. After all, finding a contradiction—any contradiction—in a system is bound to be easier than providing a direct demonstration. It was a promising approach, and it was adopted by most of Saccheri's successors in their efforts to prove the parallel postulate. Ironically, it was precisely this approach, which relied on systematically imagining a world in which the parallel postulate was false, that led to the development of non-Euclidean geometry.

Saccheri's first accomplishment was to define the problem in clear terms. In a quadrilateral *ABCD* the lines *AD* and *BC* are equal in length and form right angles (α and β) with *AB* (Figure 8.2). What can one say about the angles γ and δ at the corners *C* and *D*, respectively?[9] First Saccheri showed—without relying on the parallel postulate—that γ and δ are always equal to each other. Consequently, he pointed out, there are three possibilities:

γ is equal to δ, and both are right angles that equal 90°;
γ is equal to δ, and both are obtuse angles greater than 90°;
γ is equal to δ, and both are acute angles smaller than 90°.

The first possibility seems self-evident to us because it holds true in standard Euclidean geometry. Saccheri called it the *hypothesis of the right angle* (HRA). The other two strike us as odd because they contradict what we think we know about geometry. The first of these Saccheri called the *hypothesis of the obtuse angle* (HOA), and the second the *hypothesis of the acute angle* (HAA). Both of them directly contradict the parallel postulate. Saccheri's plan was to draw necessary conclusions from HOA until he reached a contradiction, and then do the same for HAA. Once both of the hypotheses denying the parallel postulate had been proved false, the only option left standing would be HRA, which holds true when the parallel postulate is in effect. Because it is the only hypothesis that does not lead to a contradiction, HRA is always true, and so is the parallel postulate.

Saccheri started by showing that for each assumption, HRA, HOA, or HAA, what holds for one quadrilateral holds for all quadrilaterals. In other words, each hypothesis defines a geometry in which the parallel

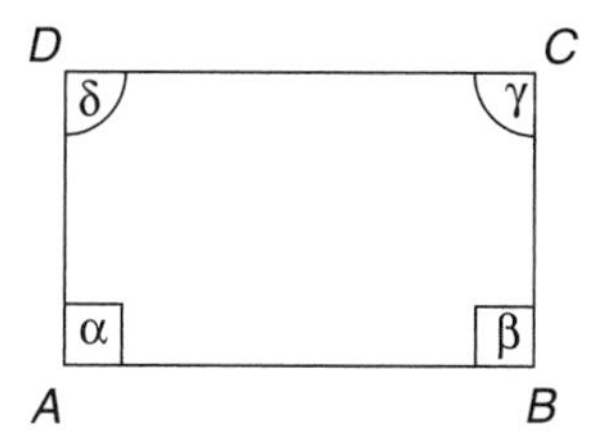

Figure 8.2. Saccheri's quadrilateral.

postulate either holds always and in all cases (HRA) or never holds at all (HOA, HAA). Then, beginning with the HOA, Saccheri showed that under this assumption the sum of the angles of a triangle is not 180°, as it always is according to Euclid, but is always greater than 180°. Furthermore, under the HOA, the sum of the angles of a triangle is not fixed but depends on the area of the triangle: the greater the area, the greater the sum of the angles. Finally, as Wallis had shown, the existence of similar figures, identical in shape but not in size, implies the parallel postulate. Saccheri now showed that this was indeed true: in a geometry in which HOA holds true, changing the size of a figure, that is, the length of its sides and its area, also changes its angles. Shrinking or enlarging a figure while retaining its shape is impossible under the HOA, and similar figures do not exist.

A world in which the HOA held sway would be a strange one indeed, but that in itself does not prove that the HOA leads to a logical contradiction, so Saccheri continued his investigations until he found one. Looking at two lines that intersect at an acute angle, say, *AB* and *AC,* he dropped perpendiculars from one to the other at regular intervals (Figure 8.3). Now in Euclidean geometry, if the intervals on *AC* (AM_1, M_1M_2, and so on) are equal, then the intervals projected on *AB* (AN_1, N_1N_2, and so on) are also equal. But under the HOA, Saccheri showed, even if the intervals on *AC* are all equal, the projected intervals on *AB* will grow larger with each step. This means that $AN_1 < N_1N_2 < N_2N_3$, and so on. Now suppose that a line *DB* intersects *AB* at a right angle. If we extend *AC* far enough, Saccheri reasoned, than at some point a perpendicular *MN* will surpass the line *DB*. This is unavoidable because as we extend *AC* indefinitely, the intervals N_iN_j on *AB* will get larger and larger and will eventually surpass any point on *AB*, no matter how far it is from *A*. And once the intervals have moved beyond *B,* that means that a perpendicular *MN* has moved beyond *BD,* and the line *AC* has necessarily crossed *BD.*

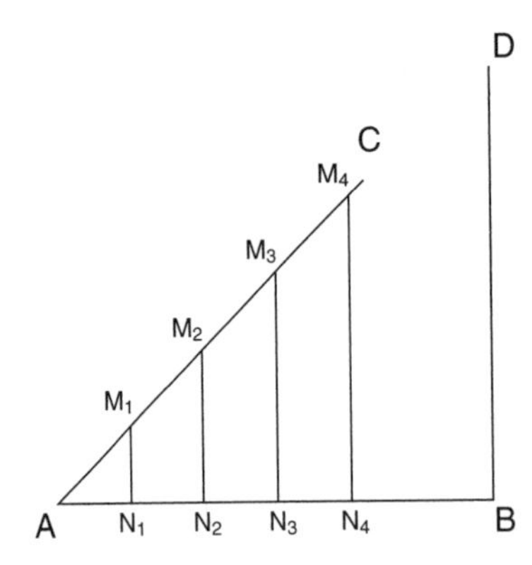

Figure 8.3. Saccheri's refutation of the hypothesis of the obtuse angle (HOA).

This shows that under the HOA, when a line *AB* intersects two other lines, *AC* and *BD*, one at an acute interior angle and the other at a right angle, the two lines will necessarily meet on the side of the acute interior angle. Saccheri then showed that when two lines intersect a third and the sum of the interior angles is less than 180°, it is always possible to draw the intersecting line so that it matches the conditions of his proof, that is, one interior angle is acute, and the other is a right angle. It follows that under the HOA, if two lines are intersected by a third and the sum of the interior angles is less than 180°, the two lines will ultimately meet on the side of these angles. This, however, is a restatement of the parallel postulate itself. In other words, Saccheri argued, under HOA we assume that the parallel postulate is false and then prove that it is true. This is a contradiction, and the HOA, Saccheri concluded, "destroys itself."[10]

Saccheri was right. If one assumes all of Euclid's postulates but replaces the parallel postulate with the HOA assumption, a logical contradiction inevitably follows. It should be noted, however, that in his proof given here, showing that the lines *MN* will ultimately surpass *BD* along the line *AB*, Saccheri relied on the fact that the line *AB* can be extended indefinitely. At some point down that line, he argued, although we do not know how far, *MN* will pass *DB*. This assumption is true in Euclidean geometry, and therefore Saccheri is correct in using it in his HOA geometry, which differs from Euclid only in respect to the parallel postulate. But what if we do away with the assumption of infinite lines and assume instead that lines are finite? In that case Saccheri's proof that HOA "destroys itself" would not hold, leaving open the possibility that an HOA geometry is in fact viable. Can we, then, imagine a world in which HOA holds true?

As it turns out, quite easily. In the years after the publication of Saccheri's *Euclides vindicatus,* Johann Heinrich Lambert (1728–1777) and later Franz Adolph Taurinus (1794–1874) pointed out that the geometry on the surface of a sphere is in fact an HOA geometry. If a segment of a sphere's great circle corresponds to a Euclidean straight line, then all lines have a fixed finite length, and Saccheri's refutation of the HOA does not hold. Furthermore, there are no parallel lines in spherical geometry because great circles always meet (twice), and the sum of the angles of a spherical triangle is always greater than 180°, the exact total depending on the triangle's area. As a result, there are no similar triangles on a sphere, and it is impossible to enlarge or shrink figures to scale while retaining their shape. All this is just as Saccheri had predicted for an HOA geometry.

Eighteenth-century mathematicians were very familiar with spherical geometry, which had played an important role in cartography and navigation for centuries. They well knew that geometrical relations are different on a sphere than they are on a plane, and none of them considered this to be a challenge to Euclidean geometry. Spherical geometry was viewed simply as the projection of Euclidean geometry on a spherical surface that itself resided in Euclidean space and could in no way substitute for it. Nevertheless, the fact that a perfectly consistent geometry that operated under such strange rules could be shown to exist on the surface of a sphere seemed to question the traditional claim that Euclidean geometry represented absolute and necessary truth.

Once he had disposed of the HOA, Saccheri turned his attention to the hypothesis of the acute angle (HAA), and here he found the going more difficult. Unlike the situation in HOA geometry, where any two lines always meet and there are no true parallels, in HAA geometry every line has not one but an infinite number of parallels. The argument Saccheri had employed to disprove the HOA did not work in this case, because in the case of the HAA the intervals *NN* along *AB* grow smaller, not larger, even as the intervals *MM* along *AC* remain fixed (see Figure 8.3). It was therefore far from clear that the perpendiculars *MN* would ever move beyond the line *BD*, even if *AC* were extended indefinitely, and it could not be proved that the lines *AC* and *BD* would ever meet. Saccheri therefore went a different route and ultimately showed to his own satisfaction that if the HAA is true, then two different lines have a common perpendicular at the same point. This, he stated, "is contrary to the nature of a straight line" and is therefore false. Just as in the case of the HOA, the HAA also has been shown to "destroy itself."[11]

At this point Saccheri was satisfied that he had indeed vindicated Euclid "from all blemish." He had shown that of the three possible hypotheses, HRA, HOA, and HAA, the latter two led inescapably to logical contradictions. Only the HRA, which corresponds to Euclid's parallel postulate, was left intact. The only viable geometry, he concluded, is Euclidean geometry. But although Saccheri may have been content with his demonstration, not everyone was convinced. His proof that the HOA is self-contradictory (at least under Euclid's assumptions) was widely accepted, even if it was not often credited to its author. The proof of the falsity of the HAA, however, was not. The problem was that in Saccheri's proof the point at which two different lines have the same perpendicular lies at infinity. Translating the rules of finite distances to a hazy infinity is a dangerous exercise, and eighteenth-century mathematicians, schooled in the hazards of the infinitesimal calculus, knew this well. Like Saccheri, they too believed that Euclidean geometry was true and necessary, and that the HAA must at some point lead to a contradiction. But they were nonetheless skeptical of the Jesuit professor's reasoning and doubtful whether he had in fact shown a contradiction.

Enlightenment Geometry

Among those who in the following years commented on parallels and their implications were several leading French mathematicians. Alexis Clairaut, d'Alembert's rival at the Paris Academy, published his own *Éléments de géométrie* in 1741, which was meant to adapt Euclid to the world of the Enlightenment. Typically for his generation, Clairaut had little patience for Euclid's meticulous rigor that had been so greatly admired by generations of mathematicians. "I can be criticized, perhaps, at some points of these *Éléments,* for relying too much on the evidence of the eyes and not attaching enough importance to the rigorous exactitude of proofs," he wrote.[12] But like his contemporaries d'Alembert and Euler, Clairaut believed that an overemphasis on logical niceties could come between mathematics and its proper object of study—the rules governing the physical universe. His *Éléments* was consequently short on rigorous demonstrations but long on practical applications, designed to show how Euclidean postulates and theorems were manifested in the real world.

Clairaut's treatment of parallels is typical in this regard. Euclid had defined parallel lines by the minimal possible requirement, stating simply that they are different lines that do not meet. Clairaut, in contrast, defined parallels according to their characteristic that is most familiar from everyday experience: that they are everywhere at the same distance from each other. From the perspective of Euclidean geometry, Clairaut's definition is far more demanding and less obvious than Euclid's because it makes a claim about every point along a line, trailing off into infinity. Proving that this is equivalent to Euclid's simple definition is a far-from-trivial exercise, but for Clairaut, such concerns were the domain of pedantic bores who had become overly immersed in the details of geometry and had lost sight of its meaning and purpose. Mathematics for him, as for most of the grands géomètres of his age, was derived from our experiences of the world and in return brought us deep knowledge of the world. A definition based on our intuitive knowledge of parallels was accordingly preferable to Euclid's elegant, but dry and detached, formulation. Characteristically, Clairaut then moved on to discuss the uses of parallels in the construction of streets and canals.[13]

Because Clairaut's starting point was the world as we experience it, he took it for granted that Euclidean space was true space and, for all intents and purposes, the only possible space. Accordingly, he never doubted the truth of the parallel postulate, which plays such an important part in describing our three-dimensional world. We know from experience and intuition that it is true, and that is all there is to it. Euclid's vaunted *Elements* was of interest to Clairaut insofar as it described the world; as an elegant mathematical edifice it was impressive, no doubt, but largely irrelevant. As a result, he was not much concerned with the alleged blemishes in Euclid, and with the fact that the fifth postulate appeared so different from the others. These were aesthetic concerns that told nothing about physical reality.

A somewhat similar view was taken by Pierre-Simon Laplace, possibly the last grand géomètre of the Enlightenment. Whereas Clairaut derived the parallel postulate from experience and engineering work, Laplace derived it from the most unshakable scientific theory of his day, Newton's cosmology. Under the principle of universal gravitation, Laplace pointed out, if the masses of all the objects in the universe, their distances, and their velocities were all changed by a fixed proportion, the shape of their motions would remain exactly the same. The Newtonian universe, in other words, was scalable—it could be expanded or shrunk without changing the relative position of its bodies or their motion. Wallis and Saccheri had already shown that scaling was possible if and only if the parallel postulate held sway. Since according to Newtonian physics the physical universe is scalable, Laplace argued, the parallel postulate is necessarily true.[14]

The argument from advanced Newtonian science sets Laplace apart from Clairaut, who tried to keep his geometry as close as possible to daily experience and practical applications. But the two shared this basic assumption: the test of the parallel postulate is whether it truly and correctly describes our world. If the world is Euclidean (and neither Clairaut nor Laplace doubted that it is), then the postulate is true, and nothing more needs to be said on the subject. This was a typical Enlightenment attitude, which assumed that the object of mathematics was to describe the underlying relationships that govern the world. Eloquently expressed in d'Alembert's "Preliminary Discourse to the *Encyclopédie,*" we have seen this approach manifested in the works of d'Alembert himself, Johann Bernoulli, Euler, and Lagrange. It is hardly surprising that Clairaut and Laplace simply assumed this approach when dealing with the parallel postulate.

But farther away from Paris, the capital of the Enlightenment, different attitudes prevailed. Johann Heinrich Lambert was born in the Alsatian town of Mulhouse and spent most of his life moving between the major centers of learning in Germany. In 1764, at Euler's invitation, he became a member of the Berlin Academy and remained there until his death 13 years later. Although Lambert is not as famous as his older contemporaries Euler and d'Alembert, he was acknowledged in his time as a leading member of the Republic of Letters, was a friend and correspondent of the philosopher Immanuel Kant, and published extensively on natural philosophy, as well as mathematics. He is best known to later generations for his theory that the Milky Way is a disk-shaped assembly of innumerable stars, one of many in the universe, and for his proof that the number π is irrational.[15] But he also devoted a book titled *Theorie der Parallellinien* to the parallel postulate.[16]

Lambert followed Saccheri in exploring the HOA and the HAA, and although his definitions differed slightly from those of Saccheri, they were ultimately equivalent to them.[17] Like his Italian predecessor, Lambert showed that the HOA leads to a contradiction as long as one assumes the infinitude of lines. He then turned to the HAA and, like Saccheri, found it much harder to identify a contradiction. He showed that the sum of the angles of an HAA triangle is less than 180°, and that it decreases as the area of the triangle increases. He then went further than his predecessors by stating—correctly, though without proof—that the area of a triangle with angles α, β, and γ is proportional to the expression $180° - (\alpha + \beta + \gamma)$.

This led him to an insightful observation. He noted that the area of a triangle on a sphere with radius r is $r^2[(\alpha + \beta + \gamma) - 180°]$. But if $r^2 = -1$, we get precisely the expression for the area of a triangle under HAA, $180° - (\alpha + \beta + \gamma)$. This would happen, however, only if $r = \sqrt{-1}$, which would make the radius r an imaginary number. "From this I should almost conclude," Lambert wrote, "that the third hypothesis [the HAA] would occur in the case of the imaginary sphere."[18] In other words, the HAA would hold true on the surface of a sphere with an imaginary radius. Lambert was intrigued by this discovery, but in the end he could not follow it up. Neither he nor any of his contemporaries had any idea what an "imaginary sphere" was, whether it could exist, or what it looked like.

In the end, Lambert left the question of the truth of the HAA in abeyance. He did not manage to prove it false, not even to his own satisfaction, as Saccheri had done. At the same time it directly contradicts our experience, and so it clearly isnt true. It might describe relations on a mysterious "imaginary sphere" in the same way that the HOA describes the relations on the "real" sphere, but whether the mathematical concept of an imaginary sphere described an actual object remained an open question. At this point, having failed to prove the parallel postulate or reach any definite conclusion, Lambert gave up and focused his attention on questions that promised more immediate returns. His *Theorie der Parallellinien* was published only in 1786, nine years after his death.

Nevertheless, although Lambert was apparently dissatisfied with his results and never published them, his work marks a subtle but crucial shift in the way the problem was perceived. Saccheri, as we have seen, was motivated by a desire to prove Euclid infallible and Euclidean geometry a worthy model for true knowledge. His French successors were interested less in geometry's logical coherence and more in the fact that it correctly described the world we live in. But all of them, regardless of their motivations, never doubted that the parallel postulate was true and necessary and correctly described our world. The alternatives were clearly false, and all that remained was to point out their internal inconsistencies. This, admittedly, had proved a more difficult task than expected, but that did not change the fundamental truth: the parallel postulate is obviously true, and it is the only one that can describe the world. The alternatives implied strange and monstrous geometries and are obviously false.

Lambert, in contrast, took nothing for granted but let his train of deductions take him where it would. Since the assumption of the HOA led to a contradiction, he concluded that the HOA was false (at least if one accepted the infinitude of lines). But since the HAA led to no discernible contradiction, Lambert was unwilling to dismiss it out of hand. Saccheri in similar circumstances forced a vague proof, which he claimed showed that the HAA was "contrary to the nature of a straight line." Lambert, in contrast, began to speculate on what an HAA world would look like. He deduced what the precise relationship between the area of a triangle and the sum of its angles would be in such a world, and from this relationship he concluded that the HAA would hold on a sphere with radius $\sqrt{-1}$. Admittedly, Lambert could make little sense of this result and ended up abandoning his investigations. Nevertheless, more than any of his predecessors and contemporaries, Lambert was willing to consider a world in which a geometry other than that of Euclid would hold true.

By seriously considering the geometry of a world radically different from ours, Lambert was charting a new course. However tentatively, he was proposing that mathematics could be pursued separately from physical reality. If a non-Euclidean world could be a legitimate object of mathematical speculation, then mathematics was not dependent on our world for its subject matter. It is possible, according to Lambert, that a world different from our own, but logically consistent in itself, is a legitimate object for mathematical speculation. Mathematical truth, in other words, need not refer to worldly reality; perhaps it need only refer to itself and rely on its own internal standards of logic and consistency.

It is difficult to say why Lambert was willing to challenge the prevailing view of the nature of mathematics while his contemporaries were not. Perhaps it had something to do with the fact that Lambert pursued his career in Germany and not in Paris, the center of European learning, where the ascendant doctrine of worldly mathematics went unchallenged. Certainly in the following decades French mathematicians proved largely uninterested in the development of non-Euclidean geometries, while German mathematicians and those in their intellectual sphere took the lead in the field.[19] Perhaps it had something to do with Lambert's enthusiastic belief in the plurality of worlds, the doctrine that the universe is sprinkled with innumerable worlds inhabited by intelligent beings. After all, one could not rule out the possibility that alien worlds bearing alien life would also be governed by alien geometries.[20] But whatever the sources of Lambert's unorthodoxy, it is undeniable that the next mathematician to explore seriously the possibility of non-Euclidean geometry shared both of these traits with Lambert. He was German, a strong supporter of the plurality of worlds, and widely recognized as the greatest mathematician of his age outside France.[21] He was Carl Friedrich Gauss (1777–1855), professor of astronomy at the University of Göttingen and known already in his own lifetime as "princeps mathematicorum"—the prince of mathematicians.

The Prince and the Paupers

Some historians of mathematics consider Gauss the true founder of non-Euclidean geometry. As his letters reveal, Gauss had indeed come to the conclusion that Euclidean geometry was not the necessary and infallible construct that generations of mathematicians had assumed that it was. "I am becoming more and more convinced," he wrote to his former student Christian Ludwig Gerling (1788–1864) in 1817, "that the necessity of our geometry cannot be proved." Geometry, he went on, should not be placed alongside arithmetic, which is a priori and derived from reason itself, "but rather in the same rank as mechanics," which is an empirical science based on experience.[22] In other words, Gauss was suggesting that while Euclidean geometry does describe space as we know it, this is because it is based on the empirical knowledge provided by our senses of how the world happens to be structured. It is not because Euclidean geometry is necessarily and exclusively true; in fact, he was implying, it may be possible to find other geometries, just as mathematically valid, whose only drawback is that they do not conform to the evidence of our senses.

This is indeed an important break with the Euclidean tradition. By downgrading the status of geometry from a priori truth to an empirical science, Gauss was challenging the field's traditional role as a cornerstone of true knowledge. But it is not clear from this whether Gauss thought that alternative geometries were, in fact, a worthy subject for mathematical investigation. After all, one might agree that mechanics is an empirical science, and that alternative principles of mechanics might hold sway in a different reality, and yet have no interest in exploring these otherworldly sciences that have no bearing on our world. If geometry is like mechanics, as Gauss suggested, it may well follow that while other geometries are possible, they are of little interest or relevance to us. We are, for all practical purposes, trapped in a Euclidean world.

In his seeming ambivalence toward alternative geometries, Gauss is a transitional figure between attitudes that prevailed during the Enlightenment and those that dominated later in the nineteenth century. In his willingness to consider non-Euclidean worlds, Gauss can be viewed as a forerunner of the view that all geometries, however alien, are mathematically valid as long as they are logically consistent. But his framing of the question looks back to eighteenth-century traditions that placed the physical world at the center of mathematical reasoning. Typically, Gauss asked not what kinds of logically valid geometries one can create, but rather how Euclidean geometry relates to the world. Unlike Saccheri, Gauss boldly concluded that our familiar geometry is not logically necessary, but simply descriptive of reality as we know it. Nevertheless, for him, as for his Enlightenment predecessors, the ultimate test of mathematics is its application to the physical world.

Gauss was ambivalent about his results in another respect as well: although we know from his papers and correspondence that his study of the parallel postulate and its implications spanned several decades, he never published a single line of his own work on the topic. Why this was so has remained a source of speculation ever since. Some have pointed to Gauss's motto, "Pauca sed matura" (few but ripe), which referred to Gauss's preference for publishing only highly polished and perfected pieces rather than works in progress. In his own correspondence Gauss referred to his fear of being subjected to "the clamour of the Boeotians" and being pilloried by his peers if he publicly endorsed such heresies.[23] It is also possible, however, that Gauss's reluctance to publish was a reflection of his own ambivalence toward the status of geometry. It is one thing to speculate boldly among friends about the nature of geometry and its relation to the world; it is quite another to publicly undermine the foundation of mathematics' claim to true knowledge without being able to offer an alternative.

Early in his mathematical career Gauss composed two memoranda on the theory of parallels, in which he expanded on the work of Saccheri and Lambert. But most of what we know about Gauss's mature thought on the topic comes from his prolific correspondence, which extended throughout the European scientific world. Without actually having published anything on the topic himself, Gauss became the acknowledged arbiter of the quality and worthiness of work in the field. It seems that everyone who wanted to contribute to the discussion on the theory of parallels wrote to Gauss and anxiously sought his good opinion. As a result, it is often difficult to say where Gauss's original contributions end and those of his correspondents begin.[24]

One of Gauss's earliest correspondents on the topic was his friend from his student days in Göttingen, the Hungarian nobleman Farkas (Wolfgang) Bolyai. Farkas Bolyai had devoted years to the study of the parallel postulate and had become convinced that he could prove the fifth postulate from simple Euclidean premises. In 1804 he sent his work to his old friend, who had now become one of the leading mathematicians in Europe. Gauss was skeptical because, as he wrote to Bolyai, he had by this time come to seriously doubt whether the parallel postulate could be proved. Sure enough, he did find an error in Bolyai's proof, and the discouraged Hungarian never again tried his hand at resolving the recalcitrant problem.

Gauss seems to have been more impressed by the work of his former student Friedrich Ludwig Wachter (1792–1817), who wrote to him from Danzig in 1817.[25] And he was truly delighted when the following year he heard from another former student, Christian Ludwig Gerling of the University of Marburg, about the work of the professor of jurisprudence Ferdinand Karl Schweikart (1780–1859). "The letter of Herr Professor Schweikart has given me extraordinary pleasure," he wrote to Gerling, "and I would really like to say a lot of good things to him about the work."[26] In his memorandum Schweikart developed what he called an "astral geometry," which was based on the denial of the fifth postulate and corresponded closely to Lambert's and Saccheri's HAA geometry. In addition to demonstrating once again that the sum of the angles of a triangle in astral geometry is less than 180°, and that this sum decreases as the area of the triangle is increased, Schweikart added this surprising result: "That the altitude of an isosceles right-angled triangle continually grows as the sides increase, but it can never become greater than a certain length, which I call the *Constant.*" Euclidean geometry holds true, he continued, if this constant is infinite.[27]

What is startling in Schweikart's memorandum is not just his results, but his matter-of-fact assertion that his new construct is a perfectly legitimate geometry that can exist alongside Euclidean geometry. "There are two kinds of geometry," he stated in the opening of his memorandum. "A geometry in the strict sense—the Euclidean; and an astral geometry."[28] He even speculated that astral geometry could potentially be the true geometry of the universe. As long as the constant was large enough, we would not notice, since for all practical purposes we would still be living in a Euclidean world. This bold assertion that an alternative geometry was not only possible but could even exist in our world should mark Schweikart as one of the founders of the field of non-Euclidean geometry. But like Gauss, Schweikart never published his results, and his work was only recovered decades later with the publication of Gauss's own correspondence.

Schweikart passed on his interest in alternative geometries to his nephew, Franz Adolph Taurinus, who also tried to interest Gauss in his investigations. Like Lambert, Taurinus suspected that a non-Euclidean geometry might hold true on the surface of a sphere with an imaginary radius. He therefore systematically translated the trigonometric equations of spherical geometry into similar equations of what he called log-spherical geometry by substituting the spherical radius r with the imaginary radius ir $(i = \sqrt{-1})$. In this way Taurinus developed a self-consistent alternative trigonometry, in which the sum of the angles of a triangle is always less than 180° and grows smaller as the area of the triangle increases.[29] But despite his success in elaborating this hypothetical trigonometry, Taurinus remained convinced that the only true geometry was that of Euclid. The log-spherical trigonometry was theoretically interesting, but in the end it was just a set of algebraic relations that could never be manifested in space. For Taurinus, as for most of his predecessors, the true test of geometry was its correspondence to the real world, and in this regard Euclidean geometry stood unchallenged.

Gauss reacted favorably to Taurinus's work, sharing with him some of his own thoughts on the subject but asking him to keep Gauss's unorthodox views to himself. Unlike his uncle, Taurinus did publish his views on the topic in two treatises dating from 1824 and 1825, but these made little impression on his contemporaries. Like Schweikart, Taurinus's reputation had to wait until Gauss's correspondence was published later in the century, and uncle and nephew were acknowledged in retrospect as forerunners of non-Euclidean geometry. Gauss meanwhile remained comfortably above the fray, refraining from any public pronouncements on the topic while privately claiming credit for radical innovations.

"Either Caesar or Nothing"

So things stood when in 1832 Gauss received another missive from his old friend Farkas Bolyai. By this time Bolyai had for several decades been living in his family home in Transylvania while serving as a professor of mathematics at the local Evangelical Reformed College of Marosvásárhely. Despite his setback in trying to prove the parallel postulate, Bolyai had not neglected his mathematical work, and in his letter to Gauss he announced the publication of his magnum opus, the result of years of intense labor: a two-volume work in Latin on the foundations of geometry, algebra, and analysis with the impressive title *Tentamen juventutem studiosam in elementa matheseos puræ elementaris ac sublimioris methods intuitiva evidentiaque huic propria introducendi, cum appendice triplici.*[30] Also included in the letter was a preprint of an appendix to the *Tentamen* by Farkas's young son János on the very question that had occupied the father years before—the parallel postulate. It was the elder Bolyai's hope that his son's treatment of the thorny problem would elicit a more favorable response from the great master than his own work had years before.

The elder Bolyai had not always been so supportive of his son's work on the parallel postulate. His own failed attempts to resolve the question had left him deeply shaken, and when in 1820 János first wrote to tell him of his work, he responded vehemently. "It is unbelievable that this stubborn darkness, this eternal eclipse, this flaw in geometry, this eternal cloud on virgin truth can be endured," he thundered in a letter to his son.[31] Nevertheless, he entreated, "You must not attempt this approach to parallels. I know this way to the very end. I have traversed this bottomless night, which extinguished all light and joy in my life." To avoid the same fate, he admonished his son, he must shy away from the problem as from "lewd intercourse." Otherwise, Farkas warned, "It can deprive you of your leisure, your health, your peace of mind, and your entire happiness."[32] Giving vent to years of pain and pent-up frustration, he continued:

> I thought I would sacrifice myself for the sake of the truth. I was ready to become a martyr who would remove the flaw from geometry and return it purified to mankind. I accomplished monstrous enormous labors . . . I turned back when I saw that no man can

> reach the bottom of this night. I turned back unconsoled, pitying myself and all mankind. Learn from my example: I wanted to know about parallels, I remain ignorant, this has taken all the flowers of my life and all my time from me.

When János would not give up his quest for the parallels, Farkas tried again to discourage his son, telling him that he expected "nothing" from his work:

> It seems to me that I have been in these regions. That I have traveled past all reefs of this infernal Dead Sea and have always come back with broken mast and torn sail. The ruin of my disposition and my fall date to this time. I thoughtlessly risked my life and happiness—*aut Caesar aut nihil* (either Caesar or nothing).[33]

Young János, however, was not a youth to be easily swayed by his father's pleas. In 1820 he was an 18-year-old cadet in the imperial engineering academy in Vienna, and two years later he enlisted in the Austrian army as a sublieutenant in the engineering corps. The handsome Hungarian aristocrat cut a dashing figure as a young officer, famous among his peers for his skills as a dancer, as well as for his fiery temper and outstanding swordsmanship. According to one story he was once challenged to a duel by 13 opponents and agreed to fight them all on condition that he be permitted to play the violin between each encounter. This being granted, he proceeded to dispatch all 13 in succession.[34] Reckless as he was in his personal life, it is perhaps less than surprising that young János ignored Farkas's fatherly advice. Throwing caution to the wind, he launched with abandon into the "bottomless night" of the parallels, and by 1823 he felt confident enough of his impending success to write to his father:

> I am determined to publish a work on the parallels as soon as I can put it in order, complete it, and the opportunity arises. I have not yet made the discovery, but the path that I am following is almost certain to lead me to my goal, provided this goal is possible. I do not yet have it, but I have found things so magnificent that I was astounded . . . All I can say now is that I have created a new and different world out of nothing.[35]

Indeed he had. For in addressing the thorny question of the parallels, János Bolyai had discovered a new mathematical universe in which our familiar laws of space do not hold and are replaced by a strange and wonderful reality. Some of his predecessors, including Gauss, Schweikart, Taurinus, and others, had glimpsed parts of this alternative universe, though they differed in their interpretations of what they saw. But Bolyai, along with his Russian contemporary Nikolai Lobachevskii, was the first to provide a complete and coherent geometry of this "new and different world."

Gauss, like most of Bolyai's predecessors, had referred his mathematical questions ultimately to the universe we live in, and was not much interested in fantastic new geometries for their own sake. His interest was in the light that the existence of such strange creations shed on the status of the one true geometry—the Euclidean—in our physical world. In particular, he reasoned, if a consistent non-Euclidean geometry was possible, then Euclidean geometry could not be considered a form of a priori knowledge but was on a par with the empirical sciences.[36]

Young Bolyai's attitude was different. Leaving our Euclidean world behind, he proceeded to actively create an alternative universe as true and real as our own. Step by step he formulated the geometry and trigonometry of his new world, establishing the basic theorems that governed it. Then, to top it off, Bolyai showed how the new geometry allowed one to accomplish something that is impossible in our Euclidean world—squaring the circle, that is, constructing a square whose area is equal to that of a given circle. For mathematicians since ancient times, and even for the broader public, "squaring the circle" had become a common expression for a hopeless quest for something impossible. By showing that in his non-Euclidean world this feat could in fact be accomplished, Bolyai was indicating that the quest was finally at an end, and that his new universe was in some respects superior to our own. Gauss, even when venturing into the strange realms of alternative geometries, had kept his gaze fixed steadily backward, on our own Euclidean reality; Bolyai, in contrast, plunged boldly into the non-Euclidean universe, reveling in its strangeness and the radical possibilities it offered.

Bolyai's determination to create "a new and different world out of nothing" poetically captures the dreams of an entire generation of early nineteenth-century mathematicians. Cauchy, Abel, Galois, and Jacobi, each in his own field and in his own way, strove to cast off the bonds that tied mathematics to our physical universe. In its place they created an alternative universe, pure, logical, and rigorous—a land of beauty governed by mathematical truth alone. Cauchy did so by reformulating the foundations of analysis and establishing the field as a self-contained mathematical system independent of physical reality. Galois did so by transforming a classic question about the solution of the quintic equation into an investigation of the deep structure of equations in general, a study that by its very nature could yield no useful results. But no one did more to establish mathematics as its own self-contained realm than János Bolyai, who announced his intention to create "a new and different world" and promptly delivered on his promise.

Like the other "new mathematicians" of his generation, Bolyai was a dramatic representative of the new mathematical persona. For if mathematics is a self-contained world of truth and beauty, then those who seek it out are brave and true souls, willing to risk all for a glimpse of this pure realm. We have already seen how Galois and Cauchy had viewed and presented themselves as martyrs to this quest, each paying a heavy personal price for his relentless pursuit of truth. And although Abel never viewed himself in such a dramatic light, the legend that grew up after his death nevertheless cast him as a tragic hero and Galois' spiritual twin. As for the Bolyais, there is no doubt that both Farkas and János, father and son, saw themselves in precisely these terms, as romantic heroes on a quest for sublime truth. The legend that surrounded their names after their deaths only added luster to the tragic personas they had themselves cultivated.

Farkas Bolyai's attempts to dispel the "stubborn darkness that stood between humanity and "virgin truth" by proving the parallel postulate ended in disaster with Gauss's harsh verdict on his efforts in 1804. Despite this, however, the elder Bolyai's spiritual quest for truth and beauty continued throughout his life. During his long years in Transylvania he turned to music and poetry, translated into Hungarian the works of romantic English and German poets, and authored five tragedies himself. This growing interest in the arts did not imply an abandonment of mathematics; to the contrary, it was an extension of the same pursuit. "The poetic turn of his spirit and the extraordinary vivacity of his imagination showed themselves in a striking manner even in his mathematical work," wrote Franz Schmidt, who authored a brief biography of the elder Bolyai in 1868. Indeed, according to Schmidt, his ability to look beyond the imperfections of the world toward a higher reality stood him in good stead not only in his artistic and mathematical pursuits but also during the troubles that descended on him during the revolutionary years of 1848–1849. Faced with devastation and civil war, he nevertheless "glimpsed beyond the harrowing spectacle of suffering imposed by human folly, a hope in the mists of eternity." "He regarded the Earth as a bog where the chained spirit languished," Schmidt concluded, "death as a liberating angel that conducts the soul out of its captivity to happier regions."[37]

All in all, whether he was engaged in mathematics, poetry, art, or even simple self-preservation, Farkas Bolyai fashioned himself in the same manner. To himself, as well as to posthumous admirers such as Schmidt, he was always a tragic romantic hero, willing to sacrifice himself in the pursuit of truth and beauty. He was a poetic soul, a figure in the mold of a Byron or a Novalis, a Mozart or a Chopin, men cut down in their quest to bring light and beauty to benighted mankind, mired in a bog of daily existence.

More than anything, however, Farkas Bolyai was a mathematician in the mold of his contemporary mathematical poets—Cauchy, Galois, and Abel. Like them, he saw mathematics as a quest for pure unadulterated truth rather than as an abstraction from physical reality; and like them, he viewed himself as a messenger from this land of truth, working to bring light to mankind suffering under error and darkness. In this, as he readily admitted, he failed, nearly losing his sanity in a hopeless quest to prove the parallel postulate. He therefore continued his quest in more traditional artistic fields such as poetry and music. But if Farkas's pursuits varied, his yearning for a pure land of truth was always the same, and it shaped his mathematical work, just as it did his poetic creations.

Nevertheless, the elder Bolyai did differ from mathematical poets of his time in one crucial respect. Galois, Abel, and Cauchy had all succeeded in their mathematical quests, at least to their own satisfaction, only to be ultimately ignored by an uncaring world. Farkas Bolyai, in contrast, never completed his mathematical journey, blazing a promising trail but turning back before possessing his great prize. In this sense the legend of Farkas Bolyai remained unfinished, for he never lived the full tragedy of the mathematical poet, the romantic genius driven to despair by a crass and uncomprehending world. These final chapters of the tragedy of the mathematical genius were left to his son, who, unlike his father, did indeed succeed in cracking the riddle of the parallels and reaching the land of pure mathematics.

That young János was a romantic like his father is clear from his heady pronouncements in the days when he was absorbed in his discovery. "I have found things so magnificent that I was astounded by them," he wrote his father, and then, in a burst of youthful hubris, "I have created a new and different world out of nothing." That he indeed succeeded in reaching this magnificent world is also beyond question, as we shall see when discussing his mathematical achievement. That he held high hopes for his future as the discoverer of this "new and different world" was evident when he wrote to his father in 1823 that although he had not completed his researches yet, "I am as convinced now that it will bring me no less honor, as if I had already discovered it."[38] Finally, that this romantic and optimistic young genius was on a collision course with the harsh realities of our less-than-perfect world is evident in what happened next.

A Mathematical Tragedy

When Gauss received the preprint of János's appendix in early 1832, he took his time and read it just as he had the works of Wachter, Schweikart, and Taurinus. Then, on March 6, 1832, he replied at length to Farkas: "If I commenced by saying that I am unable to praise this work you would certainly be surprised for a moment. But I cannot say otherwise. To praise it would be to praise myself." Following this startling assertion, Gauss then proceeded to take credit for the entire enterprise:

> Indeed, the whole contents of the work, the path taken by your son, the results to which he is led, coincide almost entirely with my meditations, which have occupied my mind partly for the last thirty or thirty-five years. So I remained quite stupefied. So far as my own work is concerned, of which up till now I have put little on paper, my intention was not to let it be published during my lifetime . . . It was my idea to write down all this later so that at least it should not perish with me. It is therefore a pleasant surprise for me that I am spared the trouble, and am very glad that it is just the son of my old friend, who takes the precedence of me in such a remarkable manner.[39]

Gauss's response to the Bolyais was typical of the man. Unwilling to risk his reputation by backing a controversial doctrine such as non-Euclidean geometries, he refrained from publishing anything on the topic and in fact, as he stated, wrote little down. At the same time, making use of his position at the apex of European mathematics, he deftly positioned himself to claim credit for the discovery should it prove profitable to do so. He kept abreast of developments in the field by corresponding with more adventurous souls such as the Bolyais, Wachter, Schweikart,

Taurinus, and Lobachevskii, assuring them of his interest in their investigations and his support. But he did nothing to promote either the work or the men and was content to let them languish in mathematical obscurity. If at some point non-Euclidean geometries would become mathematically respectable and gain the recognition they deserved, Gauss could then claim priority over all his unsung rivals as the true founder of the field. He could do this, furthermore, by making use of his reputation alone, without being required to produce much in the way of publications or even organized writings.

Some years earlier Gauss had reacted in exactly the same way to Abel's work on elliptic functions. Upon reading Abel's first article on the topic in Crelle's journal in 1827, he wrote to Friedrich Wilhelm Bessel, director of the Königsberg Observatory:

> I have not been able to complete a compilation of my own research on the elliptic functions, carried out since 1798, since I must clear away many other matters beforehand. Mr. Abel has, as I have seen, overtaken me and freed me of about one third of that labor, especially since he makes all of his development with elegance and clarity. He has chosen precisely the same path that I set upon in 1798, so the great agreement in the results is not to be wondered at. But to my surprise this agreement extends to and includes the form . . . so that many of his formulas seem to be *pure copies* of my own.[40]

As with Bolyai five years later, Gauss was complimentary of Abel's work while at the same time claiming the entire achievement as his own. Gauss furthermore made no effort to contact Abel or promote him in any way to his colleagues, leaving the young mathematician to struggle in Christiania. Abel, for his part, having spent months in Paris trying to engage the leading French mathematicians, seems to have expected little from their German counterpart. Upon leaving Paris in the fall of 1826, he wrote to Holmboe that he might pass through Göttingen "primarily to beleaguer Gauss, if he is not too strongly fortified by his arrogance."[41] In the event Abel gave up the siege and did not attempt to contact Gauss during his travels.

János Bolyai was perhaps more fortunate in his dealings in that he did see Gauss's comments on his work, whereas Abel never received so much as a single line of acknowledgment. Farkas Bolyai, furthermore, was quite satisfied with Gauss's response to his son's work and was even flattered that it corresponded closely to the great master's own efforts in the field. After all, there is no doubt that Gauss's letter of 1832 was far more favorable than his reaction to Farkas's work nearly three decades earlier. But young János was not so easily mollified: he saw Gauss's letter as a brazen attempt by the older mathematician to claim the rights to the wondrous world that he himself had "created out of nothing." Judging by Gauss's track record, it would seem that János had good grounds for his suspicions. Indeed, in the following years Gauss did nothing to promote non-Euclidean geometries, nor did he take an interest in the fortunes of the young Hungarian.

János was devastated. He quarreled with his father, whom he accused of taking Gauss's side, and even charged him with having disclosed János's discoveries to Gauss before the publication of the appendix. Father and son did not speak for several years, and their relationship remained strained for the rest of their lives. In 1833 János retired from the Austrian army as a semi-invalid, although there is no record whether this was due to mental stress over his dealings with Gauss or to the aftereffects of his dueling career. He settled, like his father, in the family seat near Marosvásárhely in Transylvania, where he lived a secluded life with his mistress. In those later years, according to Schmidt, he was known to his neighbors as a "bizarre character" who "lived in retirement from the commerce of men."[42] Although he continued to work sporadically in mathematics, he never again made lasting contributions to the field.[43] He died in 1860 at the age of 58, lonely, heartbroken, and still awaiting recognition for his discovery of a "new and different world."

For János Bolyai, the life cycle of the tragic mathematician was now complete. As a spirited young genius, he had gained entry into a wondrous mathematical world that none before him had glimpsed. Intoxicated by the thrill of his discovery, he sought to bestow it on his fellow mathematicians and indeed, as his father wrote, on all mankind. Fully aware of the earthshaking implications of his work, the young officer had no doubt that it would be welcomed by a world thirsty for truth, and that he himself would be rewarded with honor and fame. But it was not to be. When János presented his work to the leading mathematician in Germany, he was met not with the acclaim he expected but with petty jealousy and suspicion. The great Gauss was more concerned with protecting his reputation and securing his priority claims than with rewarding genius and spreading the true gospel of non-Euclidean geometry. Shocked by this encounter with the all-too-worldly side of the mathematical community, Bolyai withdrew to nurse his wounds in the isolation of his family estate in Transylvania. Although he lived for nearly 30 years after his rebuff by Gauss, his spirit was crushed, and his creativity was at an end. The eccentric Hungarian aristocrat János Bolyai died in 1860, but the mathematical genius János Bolyai had died long before, in 1832, when he received Gauss's cool response to the appendix. Coincidentally, even eerily, 1832 was also the year in which another young genius, Évariste Galois, gave up his fight for mathematical recognition and gave up his life on an empty Paris street.

The parallels between the legends of the three contemporaries—Galois, Abel, and Bolyai—are too striking to ignore. All three started their lives as mathematical prodigies, and all made profound mathematical discoveries at a very young age. Each attempted to engage the leading established mathematicians in their work, fully expecting to gain the recognition they knew well they deserved: Galois and Abel approached Cauchy, Bolyai approached Gauss. The response, in all cases, was much the same: to the horror and dismay of the young geniuses, the great geometers proved far more interested in protecting their privileged positions than in promoting either talent or truth. Tragically, the pettiness of the greatest minds of the age snuffed out the spark of genius in their innocent young colleagues. Galois died at 20, Abel at 26, and Bolyai lived out his sentence of bitterness and seclusion for three more decades but was forever lost to mathematics.

So go the legends of Galois, Abel, and Bolyai, and whether they are literally true matters less than the fact that they were told and retold, becoming in the end models of the true mathematical life. Galois, as we have seen, was the author of his own legend, as well as the source of much of his own misery. Abel never viewed himself in tragic terms at all, but he was posthumously canonized as a misunderstood and unappreciated mathematical saint. Bolyai was undoubtedly an ardent romantic, and his biography fits the tragic mold even better, perhaps, than the life stories of his two contemporaries. Both the breathless exuberance of his early letters and his victimization by the self-interested Gauss conform well to the tale of the mathematical poet. In all three cases, however, the tragic legend of the mathematical poet quickly attached itself to the memory of the young mathematicians, transforming complex life stories into simple morality tales.

Along with the poetic persona and tragic biography arrived a new approach to mathematics, which over the following decades came to

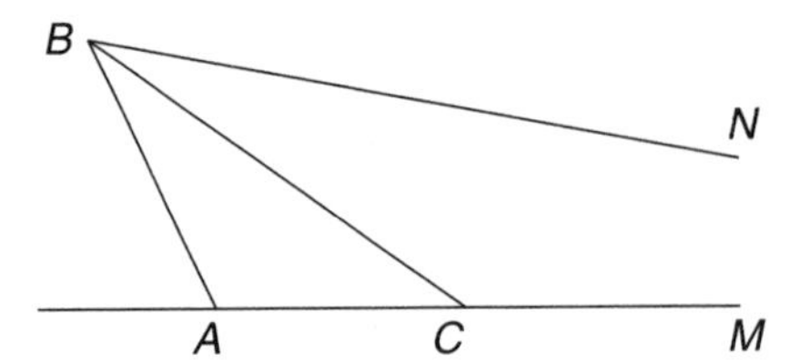

Figure 8.4. János Bolyai's parallels. After Jeremy J. Gray, *János Bolyai, Non-Euclidean Geometry, and the Nature of Space* (Cambridge, MA: Burndy Library Publications, 2006), 55, figure 4.

dominate the academic field. Galois, Abel, Cauchy, and Bolyai all believed in mathematics for mathematics' sake, independent of physical manifestations and answerable only to its own internal standards of logic and rigor. This faith in a self-contained mathematical reality found expression in their work in a variety of fields, from analysis (Cauchy) to algebra (Galois and Abel) and geometry (Bolyai). But nowhere was it more clearly and startlingly expressed than in Bolyai's detailed account of an alternative universe governed by a non-Euclidean geometry. Bolyai, along with Lobachevskii, had created a world that was mathematically flawless but conflicted directly with everything we have known and experienced about our physical surroundings. Mathematics, Bolyai had shown, was literally its own self-contained universe, and the field would never be the same again.[44]

Bolyai's World

What kind of world had János Bolyai created out of nothing?[45] Like his predecessors, Bolyai started with the assumption that the parallel postulate was false and then proceeded to define what a parallel would be in such a world. Suppose that two lines *AM* and *BN* on the same plane do not meet. *BN* would then be considered parallel to *AM* if all the lines within the angle *MAB* do intersect with *AM*. In other words, in Figure 8.4, if we start with line *BA* and rotate it counterclockwise around the point *B*, the parallel *BN* would be the first line that would not intersect with *AM*. Now if the parallel postulate were in effect, *BN* would also be the only line through *B* that did not intersect with *AM*, but in Bolyai's world this is not the case.[46] If we continue to rotate the line around *B* beyond *BN*, we will find an infinite number of lines through *B* that do not intersect with *AM*, but only one of them—the first—is considered a parallel.

It should be noted that according to this definition, each line has two parallels, one on each side of the line *AB*. In Figure 8.5 the left-side parallel

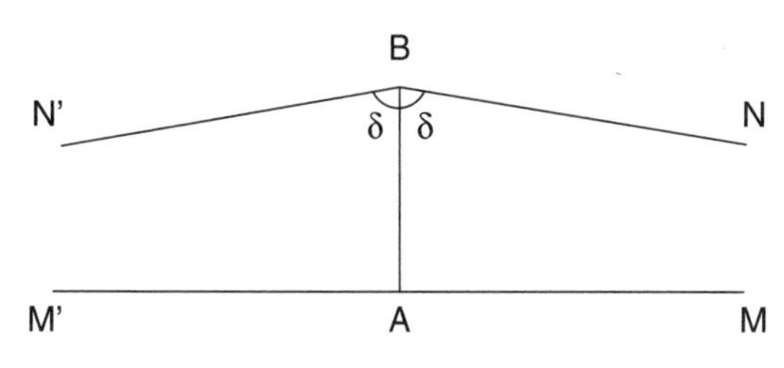

Figure 8.5. János Bolyai's two parallels. After Jeremy J. Gray, *János Bolyai, Non-Euclidean Geometry, and the Nature of Space* (Cambridge, MA: Burndy Library Publications, 2006), 56, figure 5.

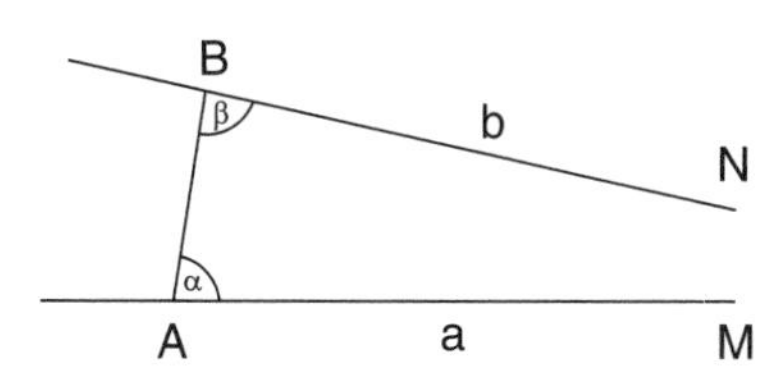

Figure 8.6. Parallels and angles in Bolyai's non-Euclidean geometry. After Jeremy J. Gray, *János Bolyai, Non-Euclidean Geometry, and the Nature of Space* (Cambridge, MA: Burndy Library Publications, 2006), 57, figure 6.

BN' would be arrived at by rotating the line *BA* clockwise around *B* and finding the first line that would not intersect with *AM* on the left. Bolyai then shows that if the angles at *A* are right angles, then the two angles of parallelism δ at *B* are equal to each other.[47] In Euclidean space δ is obviously equal to 90°, but when the parallel postulate is false, it can be either greater or smaller than a right angle. Bolyai posited that δ is smaller than 90°, an assumption equivalent to Saccheri's HAA. He then proved (Figure 8.6) that if lines *a* and *b* are parallel, and *A* is a fixed point on *a,* then there is a unique point *B* on *b* such that the angle α is equal to the angle β.[48]

From this Bolyai moved on to define a curve and a surface with unique properties. Given a line *a* and a unique point *A* on it, he considered all the lines parallel to *a* in a given plane. On each of these, according to the previous theorem, there is a unique point B such that the angles *MAB* and *NBA* are equal. In Figure 8.7 the unique point on line *b* is B, and the unique point on line *c* is C. By connecting all the points A, B, C, and so forth on all the parallels on that plane, Bolyai arrived at a rounded curve that he designated L. Nikolai Lobachevskii, working independently at the same time, also arrived at this curve and gave it the name that stuck—the *horocycle.*[49]

Bolyai then assumed that the horocycle is rotated around the original line *a,* passing through all the planes that contain the line *a* and producing a surface. In Euclidean space the horocycle would be simply a straight line, and the surface would be a plane perpendicular to the line *a.* But in Bolyai's non-Euclidean world the horocycle is a rounded curve, and the

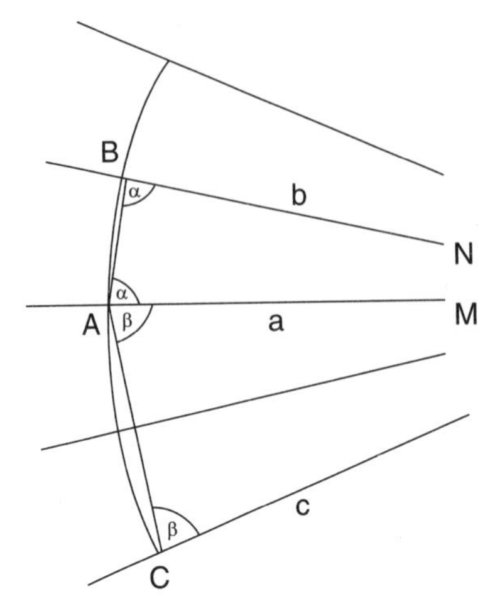

Figure 8.7. Bolyai's L curve, named by Lobachevskii the *horocycle*. After Jeremy J. Gray, *János Bolyai, Non-Euclidean Geometry, and the Nature of Space* (Cambridge, MA: Burndy Library Publications, 2006), 58, figure 7.

surface it produces is shaped like a bowl. Bolyai called this the F surface, and Lobachevskii, more memorably, the *horosphere*.

The F surface, Bolyai showed, has several interesting features in non-Euclidean space, including that it meets all the lines parallel to *a* at right angles, and that any plane containing the line *a* meets the F surface in an L curve. But his most remarkable insight was this:

> On any F surface, if two L curves cross a third and the sum of the interior angles is less than two right angles, then the two L curves intersect.

This, of course, is a formulation of the parallel postulate itself. It means that if L curves are taken to be straight lines, then the parallel postulate holds on an F surface in non-Euclidean space. In other words, just as a sphere is a non-Euclidean surface within Euclidean space, the F surface is a Euclidean surface within non-Euclidean space.

Armed with this discovery, Bolyai was now able to translate familiar Euclidean theorems into his new non-Euclidean world. In particular, he demonstrated how a triangle on the F surface, with all its familiar Euclidean properties, can be translated into an equivalent non-Euclidean triangle in the surrounding space. This non-Euclidean triangle, Bolyai showed, will preserve all of the original triangle's geometric relations. He next moved beyond simple triangles and attempted to formulate the trigonometric relations that hold in his non-Euclidean world. In order to do this, however, he had to confront the problem of the angle of parallelism,

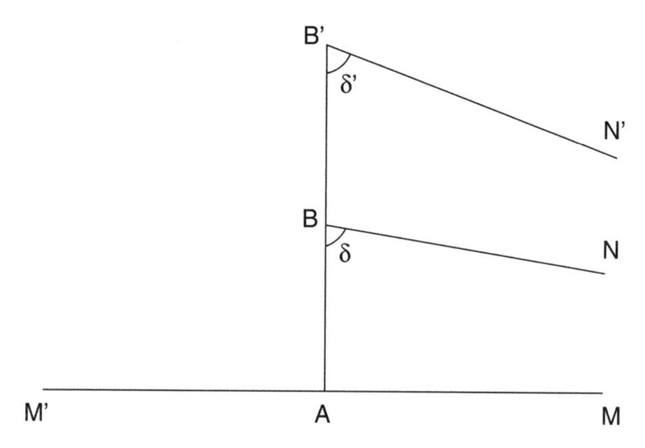

Figure 8.8. The angle of parallelism in Bolyai's geometry.

which fundamentally separates non-Euclidean from Euclidean geometry. In Figure 8.8 the segment AB is perpendicular to the line MM', and δ is the angle of parallelism. This means that δ is the smallest angle for which the line BN does not intersect with MM'. Any line inside the angle ABN will ultimately cross MM'.

Now in Euclidean geometry the angle of parallelism is always 90°. This is intuitively obvious, and it also follows directly from the parallel postulate. But in a non-Euclidean world δ is not fixed but depends on the length of the segment AB. When AB is very short, δ is very close to a right angle, or else the line BN would inevitably intersect with MM'. But if we extend AB in Figure 8.8 to AB', then the angle δ' at B' can be smaller than δ, and $B'N'$ would still not meet MM'. In other words, the longer the segment AB, the smaller the angle of parallelism δ.

In order to determine the trigonometric equations that hold in this non-Euclidean world, Bolyai first had to find the precise relationship that holds between the length of the segment AB and the angle of parallelism δ. In modern notation, if we designate the length of the segment AB as y, then the angle of parallelism δ that corresponds to y is designated as $\delta = \Pi(y)$. Following a long geometrical argument, Bolyai found that the relationship between the length y and the angle of parallelism $\Pi(y)$ is given by $\sinh(y) = \cot\Pi(y)$.[50] With this result in hand, he proceeded to deduce the formula for the circumference of a circle in a non-Euclidean world and the relationship between the sum of the angles of a triangle and its area. He followed this by his most spectacular result—the squaring of the circle.

These last results, Bolyai showed, depended on an arbitrary constant *k*, with each value of *k* producing different outcomes. In particular, when *k* was set at infinity, all the formulas reduced to their familiar Euclidean form. This inherent indeterminacy, very different from the single unchallengeable results that had so long been the pride of Euclidean geometry, had greatly troubled Farkas Bolyai when János first showed him his results. Indeed, the disagreement on the subject between father and son had likely delayed the publication of the appendix by several years. But János himself was not much concerned with this irregularity. Quite likely it seemed to him that he had discovered not one but an infinite number of "new and different worlds," each given by a different value of the constant *k*.

János Bolyai was true to his word. In 1820 he had declared that although he did "not yet have it," he was convinced that he had "created a new and different world out of nothing."[51] A dozen years later, with the publication of the appendix, he had done it. His non-Euclidean world appeared to possess everything that a Euclidean world did, and in some ways (such as the squaring of the circle) even more. All the figures and their relations familiar from Euclidean space had their counterparts in Bolyai's world. True, neither Bolyai nor any of his contemporaries had any idea what a surface or a space would look like in which Bolyai's geometry would hold true. But for Bolyai, the physical manifestations of his discovery mattered little. He had discovered a new mathematical world that was just as consistent and viable as Euclid's. Judged mathematically, there was little to choose between traditional geometry and its newly found alternatives. Both described complete, rational, and self-consistent worlds, and both were equally real. Euclid was the discoverer of one; Bolyai of the others.

The Poetic Life

János Bolyai was not the sole originator of non-Euclidean geometry. We have already noted some of his immediate predecessors, including his own father, Wachter, Schweikart, Taurinus, and Gauss himself. But more than any of his predecessors, Bolyai provided a complete and comprehensive picture of a non-Euclidean universe, and more than any of them, he understood the radical implications of his discovery: through mathematical reasoning he had discovered an alternative universe, sharply at variance with everything we know around us, but equally true. It was a notion of truth and of mathematics that went against the grain of the traditional understanding of the field, which viewed it as the abstract study of physical reality. To both practitioners and laymen, mathematics and its meaning would never be the same again.

Bolyai's only rival as the discoverer of this "new and different world" was his Russian contemporary Nikolai Ivanovich Lobachevskii (1792–1856). In the same years in which Bolyai was working out the details of what would become his appendix, Lobachevskii was developing his own version of non-Euclidean geometry from his seat as professor—later rector—at the University of Kazan. Remarkably, Lobachevskii's work parallels that of Bolyai in every respect, a fact noted by both Farkas and János Bolyai when they obtained a summary of his work in the late 1840s. Like Wachter, Schweikart, Taurinus, and the Bolyais before him, Lobachevskii also tried to interest Gauss in his work, sending him a copy of a short publication in German expounding his theories. Gauss, as usual, responded favorably and even went so far as to make Lobachevskii a corresponding member of the Göttingen Academy. Beyond this, as usual, he took no step at all, and Lobachevskii languished in obscurity in the Russian hinterland for the rest of his life.

Lobachevskii is usually mentioned alongside János Bolyai as the codiscoverer of non-Euclidean geometry. Their discoveries overlap and their life stories also converged to an extent because Lobachevskii, like young Bolyai, suffered misfortune in his career. Lobachevskii never received the recognition he deserved for his groundbreaking work, which gained broad circulation only after his and Bolyai's deaths. He served as rector at Kazan University for 22 years but was abruptly removed from his post in 1849, possibly for political reasons. He died seven years later, disgraced at home and unacknowledged abroad.

Nevertheless, in spite of these tragic final years, it is questionable whether Lobachevskii should be included among the mathematical poets of the early nineteenth century. Russian traditions, cultural narratives, and imagery are different from those of central and western Europe, and they are the ones against which the significance of Lobachevskii's career should be gauged. Did the rector of Kazan University view himself as a tragic hero? Was he viewed in this light by his Russian contemporaries? Did he view his work on non-Euclidean geometry as an effort to reach beyond our mundane surrounding to a mathematical universe of truth and beauty? Did subsequent generations canonize him as a martyr for truth? The answers to all of these questions could possibly be yes, but for the moment we do not know. A true evaluation of Lobachevskii's career will have to wait for a study that will take into account the unique cultural setting in which he lived and worked. The question whether he belongs among the tragic mathematicians of the early nineteenth century must for now remain open.[52]

János Bolyai, however, is the very embodiment of the poetic mathematician of the romantic age. A brilliant young man, an accomplished dancer and swordsman, he discovers a brave new world of truth and beauty at the tender age of 21. Eagerly he presents his discovery to one of the leading mathematicians of the age, confident that it will be recognized for the marvel that it is. But the established academician proves unworthy of the gift. More concerned with preserving his own position than with the opening of new mathematical worlds, he downplays the young genius's accomplishment and hints that he himself is the true discoverer. He then does his best to bury the whole matter and keep it from gaining broader currency. The young man is devastated: not only is he denied recognition, but the illustrious champion of pure reason has shown himself to be a petty official motivated by self-interest. Shaken to his core, the young genius retires from the scene and is never heard from again in the world of mathematics.

Such was the case of János Bolyai and his encounter with Gauss and the realities of academic politics. Such also, with minor modifications, was the story of Galois' encounter with Cauchy and the Institut, and especially the legend that grew around his name. Such was also the legend that grew around Abel after his death, although he himself had no hand in it, and such was the image that Cauchy presented of his career among his friends and admirers. Such, in other words, was the story of the tragic mathematician that appeared in the early decades of the nineteenth century, contrasting sharply with earlier depictions of geometers and their roles. In the legends and imagery that attached themselves to their names, Galois, Abel, and Bolyai are three of a kind—tragic romantic heroes who perished prematurely in their pursuit of sublime truth. And Cauchy, so often the villain in tales of mathematical persecution, was to himself and his admirers no less a martyr than his younger colleagues.

Hand in hand with the poetic mathematicians went a new poetic mathematics. Freed from the shackles of physical reality, it became a world all its own, answerable only to its internal logic and held together by rigorous mathematical reasoning. Cauchy knew it to be so when he broke with eighteenth-century practice and reformulated analysis as a self-contained rigorous system, founded on his δ-ε inequalities. Abel knew it to be so when he championed the cause of strict mathematical rigor and warned that the loose practice of his predecessors had resulted in a "vast darkness" pervading mathematics.[53] Galois knew it to be so when he devised a thoroughly impractical method for determining whether an equation was solvable by radicals. The power and beauty of the method, he claimed, lay precisely in its impracticality, and in the future, he confidently predicted, all higher mathematics would focus on such useless but profound structural problems. And more than anyone, perhaps, János Bolyai knew it to be so when he created a "new and different world out of nothing," based only on strict mathematical reasoning.

Strange Men in Strange Worlds

For more than two millennia Euclidean geometry had been the rock on which all mathematics was founded. It provided a paradigm for correct rational reasoning, capable of producing true and unchallengeable results that other fields could only envy. But just as importantly, it provided mathematics with an unshakable anchor in the physical world, connecting even the most abstract algebraic formulations to the realities we see around us. For geometry, as its name implies, has its roots in land measurement, and in its abstract Euclidean form it became the science of space, describing the hidden connections that govern our three-dimensional universe. When early modern mathematicians moved toward ever-greater algebraic abstraction, they were always careful to make sure that they had not lost contact with their geometrical roots. As d'Alembert most famously formulated the issue in the "Preliminary Discourse," geometry was an abstraction of physical reality, arithmetic was an abstraction of geometry, and algebra was an abstraction of arithmetic.[54] When all was said and done, even the most abstract algebra was about the relations that governed the physical world. Euclidean geometry was the unbreakable chain that bound the most esoteric algebraic formulas to the concrete realities of the physical world.

It was this foundational role of geometry in the edifice of mathematics that made the invention of non-Euclidean geometry such an earthshaking and traumatic affair for the entire discipline—and for rationality itself. János Bolyai and Nikolai Lobachevskii did not simply add a new field of research to mathematics, like an additional floor to an existing structure. Their discovery was far more dangerous, and indeed more subversive, than that: by challenging the necessary truth of Euclidean geometry and offering an equally valid alternative, they were undercutting the foundations of the entire house of mathematics. If geometry did not necessarily relate to the physical world, then what could be said of the rest of mathematics that was founded on it? If geometry was set free of its mooring in the physical world, then certainly algebra and analysis were unmoored as well. The house of mathematics, for thousands of years securely founded on the rock of geometry, had been set adrift and was now floating free in conceptual space.

Such was the effect of Bolyai's and Lobachevskii's invention of a new geometry—as fully coherent as that of Euclid, but at sharp variance with our experience of physical space. Along with Cauchy's refounding of the calculus, Galois' vision of a new mathematics focused on its own deep structures, and Abel's emphasis on strict rigorous deduction, it marks a turning point in the development of mathematics. Since classical times mathematics had been understood as the study of magnitude, abstracted from physical reality but nevertheless rooted in it. Now it became a pure logical construct, severed from the world and serving as its own subject matter. With physical reality no longer its criterion for truth, mathematics became a separate world, governed solely by its own standards of rigorous deduction and accessible only to the initiated few. No single mathematical development did more to make it so than the invention of non-Euclidean geometry.

Is it any wonder that this new mathematics would require a very different kind of practitioner than the traditional mathematics of Enlightenment geometers? Hardly. As long as mathematics was rooted in the physical world, its practitioners were perceived as men of the world—rooted in their natural setting and their social surroundings. Such was d'Alembert, to his admirers a natural man who never lost his childlike affinity to the world and its mysteries. Such were also his colleagues, the grands géomètres of the eighteenth century, all successful men of affairs, at home in the natural and social worlds alike.

But when mathematics in the early nineteenth century broke free of physical reality and became its own self-contained world of wonders, such practical men would no longer do. Attuned as they were to the physical world we live in, d'Alembert and his colleagues possessed no privileged access to the mysteries of an alternative mathematical reality. Such access is granted to a very different kind of person, one possessed of a special sight enabling him to look beyond the constraining circumstances of our world and into other strange and wonderful realms. If mathematics is its own "other world," then only otherworldly men are privileged to reach it.

Like the new mathematics itself, such a person does not belong to our imperfect world but to a different one, governed by pure truth and beauty. He lives among us as a stranger, misunderstood and persecuted, ending his life in disillusionment, exile, or even early death. Such is the romantic mathematical poet, who reaches out to a world of sublime perfection but is crushed by the pettiness and cruelty of our own imperfect one. Such were Galois, Abel, and Cauchy, whether in their own eyes or in those of others. And such was also János Bolyai, who as a young man had created "a new and different world out of nothing," but who lived out his years in bitter loneliness in our own all-too-familiar world.

CONCLUSION: PORTRAIT OF A MATHEMATICIAN

In 1809, 17 years before Abel arrived in Paris, another young Norwegian made the same journey in search of success in Europe's cultural capital. His name was Johan Gørbitz (1782–1853), a painter and miniaturist who at the age of 27 set out to impress the great Parisian artists of his day, just as Abel was intent on gaining recognition among the savants of the Institut years later. Unlike Abel, who was happy to return to Germany and then home at the first opportunity, Gørbitz settled in Paris and soon found a place in the workshop of Antoine-Jean Gros (1771–1835). He stayed in Paris for 26 years, becoming a moderately successful artist and displaying several of his works at the annual Salon de Paris—the official exhibition of the Paris Academy of Fine Arts. He returned to Christiania only in 1835, after his master, Gros, drowned himself in the Seine in a fit of depression over his declining artistic reputation.

It was not long after Abel arrived in Paris in the summer of 1826 that he made Gørbitz's acquaintance. Twenty years Abel's senior, Gørbitz was by then an old Parisian and was well positioned to instruct his young countryman in the ways of the city. As is often the case with expatriates in a foreign city, the two spent much time in each other's company in the long summer months when many of the people Abel had come to meet were on vacation and the libraries were closed. It was during one of these meetings that Gørbitz drew a portrait of Abel, which is the only known image of the young mathematician (Figure C.1).[1]

The image in Gørbitz's portrait is of a quietly handsome young man, well bred, serious, and thoughtful. He is comfortably dressed in a well-fitted fur coat, his hair is fair, and his mouth is set in the faintest hint of a smile, as if acknowledging the artist at work. But it is the young man's eyes that grab our attention and draw us irresistibly toward them. Dark and intense, they stand out sharply against Abel's fair complexion and

Figure C.1. Portrait of Niels Henrik Abel by Johan Gørbitz, 1826. © Department of Mathematics, University of Oslo.

Figure C.4. Évariste Galois, artist unknown, ca. 1828. First published in Paul Dupuy, "La vie d'Évariste Galois," *Annales scientifique de l'École normale superieur* 13 (1896): following page 200.

Figure C.2. Self-portrait by Alexandre Abel de Pujol (1806). Réunion des Musées Nationaux/Art Resource, NY.

Figure C.3. Self-portrait by Philipp Otto Runge (1802). Bildarchiv Preussicher Kulturbesitz/Art Resource, NY.

amiable pose. They burn with a fire that suggests deep passions of the soul and profound insights of the mind. Their gaze shoots out from the painting's surface and pierces through us, focused not on us but on a distant point on the horizon. If the initial impression of the portrait is of a polite and agreeable companion, the eyes offer a different and more profound insight: the portrait is of a man who deep down is indifferent to his present surroundings, absorbed instead by his own inner flame and a vision he perceives far beyond. Abel, in Gørbitz's portrayal, was a young man who outwardly conformed to the requirements of life in our world but—as his distant gaze revealed—was not quite of it.

Gørbitz's Abel strikes us as a singular individual, set apart from his peers by his consuming inner fire and his special gift of genius. There is some irony, therefore, in the fact that in the early nineteenth century such representations of radical individualism were quite common. They were part of the standard iconography of high romantic art, and Gørbitz, who lived and worked at the heart of the Parisian artistic milieu, knew this visual language well. His master Gros had been Jacques-Louis David's favorite pupil and had gained an enormous reputation for his heroic depictions of Napoleon on the battlefields of Arcola, Nazareth, Aboukir, and Eylau. Gros' proudest moment came in 1808, when the emperor viewed *Napoleon on the Battlefield of Eylau* at the Paris Salon and was so moved that he took his own cross of the Legion of Honor and pinned it on the artist's chest. In his later years Gros became the inspiration for a younger generation of romantic artists, who particularly admired his realistic depiction of the suffering of common soldiers.[2] As Gros' associate, Gørbitz could not fail to be intimately familiar with the work of the romantics.

When Gørbitz's depiction of Abel is compared with contemporary portraits by leading romantic artists, it becomes clear that it is of a kind with the romantics' depiction of artistic genius. Consider, for example, the self-portraits of romantic artists, such as the one by the French Alexandre Abel de Pujol (1787–1861) (Figure C.2) or the German Philipp Otto Runge (1777–1810) (Figure C.3), who died of tuberculosis at age 33. Both Runge's and Abel de Pujol's images are dominated by their dark eyes and piercing gazes that seem to look at the observer but do not see him, focused as they are on deeper and more distant truths. Runge in particular appears as a conventionally handsome and well-dressed young man whose melancholy eyes reveal a far deeper sensibility and a spark of genius.[3] The resemblance between these depictions of romantic artists and

Gørbitz's portrait of Abel is unmistakable. Evidently, the only existing portrait of the Norwegian mathematician conveys not only his individual features and demeanor but also the conventions of high romantic art.

Gørbitz's portrait is the only image of Abel that survives to this day, and although we would wish to have much more, we should, on the whole, consider ourselves lucky to have this much. The young Norwegian's portrait is a professional work executed from life by a respected artist who knew him well. Nothing approaching this quality and reliability has survived to bring before us the images of two other mathematical poets of Abel's day, Évariste Galois and János Bolyai. Of two images of Galois left to us, one is a crude sketch made from memory 16 years after his death by his brother Alfred to accompany an article in *Magasin Pittoresque*.[4] The other is an anonymous sketch taken from life when Galois was 15 or 16 (Figure C.4). It was acquired from Galois' niece by his biographer, Paul Dupuy, and it first appeared in Dupuy's *La vie d'Évariste Galois* published in 1896 by the École normale.[5] It is this portrait that has been adopted as the face of Galois in all modern publications.

Galois' portrait is much cruder than Gørbitz's drawing of Abel, but in other respects the two images are strikingly similar. Galois' portrait dates to 1827 or 1828, only a year or two after Gørbitz drew Abel during one of their meetings in the summer of 1826. The Frenchman's complexion appears rather darker than that of the Nordic Abel, but both are turned at a slight angle from the viewer and are comfortably dressed in a heavy coat. In his midteens Galois is a decade younger than Abel and shows the awkwardness of youth. He appears shyer and more reticent than the older and more confident Norwegian, as well as a touch more arrogant. But just as in Gørbitz's portrait of Abel, it is the eyes that transfix us in Galois' image. Dark and piercing, they burn with a fire that testifies to fierce passions within and reaches out to distant and profound truths. They look upon us with an ironic skepticism that belies their owner's tender years, and they convey clearly that he is not truly interested in us, who stand before him. What he sees lies far beyond our horizons.

Galois and Abel had several acquaintances in common, including Raspail and the Baron de Férussac, but they never met during Abel's sojourn in Paris. Galois had never even heard of Abel until 1830, when the Norwegian was posthumously awarded the academy's grand prize and Galois found that Abel had been working on some of the same mathematical problems that he had himself. Nevertheless, Galois' and Abel's portraits, executed only a year or two apart, exhibit an unmistakable stylistic resemblance. The iconography is nearly identical—the pose, the dress, the fiery gaze directed toward us but focused elsewhere. Above all, the underlying narrative of the two portraits is the same, telling of brilliant youths who see far beyond the mediocrity that surrounds them and do not truly belong in our mundane world. Years before their lives were to take the tragic turn that would end in their early deaths, the two young mathematicians were already portrayed as melancholy romantic dreamers.

The third of the trio of tragic mathematicians in the generation of Galois and Abel was János Bolyai, and sadly no contemporary portrait of him survives at all. According to his manuscripts he had, in a moment of despair, burned the only portrait of himself taken from life, which depicted him in the uniform of an imperial officer.[6] The portrait that is currently in wide circulation on the Internet is, unfortunately, inauthentic. It is reproduced from a postage stamp issued by the Hungarian postal service in 1960, which is itself a segment of an older inauthentic portrait. Based as it is on the imagination of an anonymous artist, the portrait tells us nothing about what young János actually looked like. But for precisely the same reason it tells us everything about Bolyai's legend, and how his memory was preserved and cultivated in his homeland. Unencumbered by any real information that might constrain his artistic expression, the artist chose to express Bolyai's myth with a portrait that closely resembles that of his contemporaries Galois and Abel, taken from life. Once again, a handsome young man looks out at us, well dressed, perhaps in a military uniform. And once again it is the piercing dark eyes that arrest our attention, gazing through us at greater things beyond. The Bolyai of legend is a tragic romantic hero in the mold of Galois and Abel, and his tragic end is already foretold in his dark gaze.

In their portraits, as in their legends, Abel, Galois, and Bolyai are misunderstood heroes whose inner fire and profound insights set them apart from their uninspired fellow men. We have already seen that this put the young mathematicians in the same class as depictions of the romantic painters of their day. However, it was not only portraits of painters but of romantic poets as well that shared the same visual iconography. An otherworldly John Keats, aged only 23, gazes dreamily into the distance in a famous portrait by his friend Joseph Severn (1818–1819) (Figure C.5). Lord Byron is in deep contemplation in Richard Westall's portrait of 1813, so fully absorbed in a reality taking place outside both the painting and the viewer's world that he appears oblivious to both (Figure C.6). In

Figure C.5. John Keats by Joseph Severn, 1818–1819. National Portrait Gallery, London.

Figure C.6. Lord Byron by Richard Westall, 1813. National Portrait Gallery, London.

either of these portraits we could easily imagine the face of a Galois, an Abel, or a Bolyai.

When the portraits of our young mathematicians are viewed side by side with those of the romantic poets of the same era, it becomes clear that the images are all of a kind, belonging to the same genre and sharing the same iconography. They speak to us as one, conveying the same narrative, of young geniuses so absorbed in pursuit of the sublime that they fall victim to the commonplace trials of life on earth. Such, as we have seen, were the legends told of romantic artists and poets, and such were also the myths that have attached themselves to the names of our tragic young mathematicians. In their portraits, as well as in their legends, Galois, Abel, and Bolyai were tragic heroes in quest of purity and truth, just as Byron and Keats sacrificed themselves in their own quest for the sublime.

But were mathematicians truly unique in this way? After all, it could be supposed, romantic iconography was so dominant in the art of the early nineteenth century that its subjects would be forced into a romantic mold regardless of their actual suitability for the role. In particular, one may wonder whether the natural scientists of the age were depicted in a manner similar to that of the young mathematicians—as tragic strivers in search of truth. A quick study of nineteenth-century images of leading scientists reveals that this was not the case. The image of the young genius so captivated by a wondrous world beyond that he is unable to cope with his immediate surroundings was, and remains, a paradigmatic image of the modern mathematician, but it is completely absent from nineteenth-century representations of leading natural scientists. The mantle of the romantic hero was reserved to artists, musicians, poets, and—as we have seen—mathematicians.

There are, to be sure, hundreds of portraits and photographs of the leading scientists of the nineteenth century, done in a range of different media and styles. A tiny sample of two well-known images cannot do justice to this numerous collection, but it can give one a taste of the conventions used in portraying a "man of science." In their best-known images the leading natural scientists of that age were far older than the iconic mathematical geniuses; they were men at the peak of their careers, whose interests and influence spanned nations and empires. They were, on occasion, presented as bookish, sometimes as eccentric, but never as tragic.

In Britain William Thomson, later Lord Kelvin, and in Germany Hermann von Helmholtz were each world leaders in their chosen fields. But they were also much more: along with their scientific research they had leading parts in major engineering and industrial ventures and became public spokesmen for "science" writ large. Consequently, their portraits reflect their active role and central position in the life of science and the nation (Figures C.7 and C.8). Lord Kelvin appears as a bearded patriarch of science, whose high brow and kind but sparkling eyes bespeak profound wisdom and deep reservoirs of experience. A somewhat younger Helmholtz is depicted among his scientific instruments. He looks directly at us, thoughtful, alert, and fully present, seemingly in the midst of a demonstration to an admiring audience. The traditional globe in the background is a reminder of the scientist's proper object of study—our own physical world. Both are successful and influential men of the world, moving in the highest echelons of society and taking a leading part in the shaping of the modern world. There is nothing about them that recalls the tragic and disillusioned youths who transformed mathematics but were destroyed by life.

It seems safe to say, then, that portraits of mathematicians took on a wholly different character than those of natural scientists in the nineteenth

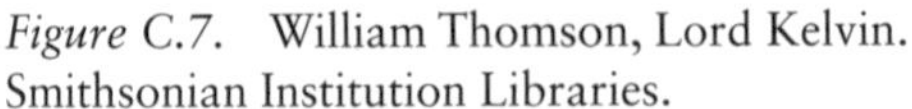

Figure C.7. William Thomson, Lord Kelvin. Smithsonian Institution Libraries.

Figure C.8. Hermann von Helmholtz. Smithsonian Institution Libraries.

century. Natural scientists were presented as engaged and worldly men who focused their minds on the real world, and whose lives were firmly anchored in the here and now. Mathematicians, in contrast, were presented as young romantic heroes, sacrificing all in their quest for purity and truth beyond the bounds of worldly existence. Like their legends and the stories associated with their lives, mathematicians' portraits diverge sharply from those of other scientists and follow instead the trends and conventions usually associated with artists, musicians, and poets.

But while the men of science of the nineteenth century are an altogether different breed from the mathematicians of their day, they do appear to be direct descendants of an earlier generation of geometers. Consider the contemporary portraits of two grands géomètres of the eighteenth century, Johann Bernoulli (Figure C.9) and Jean le Rond d'Alembert (Figure C.10). Like the natural scientists a century later, the two are depicted in the prime of life, at the height of enormously successful careers. Accomplished and energetic, they look confidently into the future, searching for new challenges that they will undoubtedly overcome. Like the natural scientists a century later, they are successful men of the world, showing

Figure C.9. Johann Bernoulli. Smithsonian Institution Libraries.

Figure C.10. Jean le Rond d'Alembert. Smithsonian Institution Libraries.

no hint of the morose sentimentality that became a hallmark of the mathematical persona in the age of Galois and Abel.

The resemblance between the geometers of the eighteenth century and the natural scientists (as they came to be called) of the nineteenth century is not, on the whole, surprising. After all, the grands géomètres of the Enlightenment were as much forerunners of the mathematical physics of Lord Kelvin and Hermann von Helmholtz as they were of the pure mathematics of Niels Henrik Abel and Évariste Galois. Never straying far from the physical roots of their mathematical abstractions, d'Alembert and his contemporaries focused their energies on questions such as the curve of a hanging chain (Johann Bernoulli) and the motion of a vibrating string (d'Alembert).

Like the men of science of the following century, the subject matter of Enlightenment geometers was the world itself, and their depictions expressed this clearly. In his portrait Johann Bernoulli is shown holding an elegant astronomical instrument, likely an astrolabe, in one hand and a geometrical diagram, seemingly derived from the instrument, in the other. Similarly, d'Alembert is in the midst of composing a geometrical tract, holding a pen in one hand and a compass in the other, with other geometrical instruments strewn on his desk. A globe is displayed prominently—and predictably—in the background. The message in both portraits is clear: the geometer's subject matter is the earth and the heavens, his method geometrical abstraction.

In their imagery, as well as in their technical practice, the new mathematicians of the early nineteenth century charted their own course, at odds with the path of their Enlightenment predecessors. The iconographic tradition that had characterized the portraits of d'Alembert and his contemporaries did not end but continued uninterrupted in the images of nineteenth-century natural scientists. But in the most enduring images of pure mathematicians this imagery was abruptly abandoned. In place of the confident and optimistic men of the Enlightenment appeared new images of young romantic dreamers, spiritual seekers alienated from our world. In the early nineteenth century the iconography of mathematics broke away from traditional representations of the natural sciences, which for millennia had been its close companions. Instead, it allied itself with the creative arts, presenting the young mathematicians of the early nineteenth century as tragic romantic heroes in the mold of a Byron, a Runge, or (later) a Van Gogh.[7]

In essence, the visual history of mathematical portraits reiterates the narrative history of mathematical stories and legends. We have already seen how the paradigmatic story of d'Alembert, the iconic natural man, was replaced by the paradigmatic story of Galois, the iconic tragic genius. Each of these stories represented far more than personal biographies of brilliant mathematicians. D'Alembert's story was emblematic of a time when mathematics was viewed as the deep science of the world, revealing hidden harmonies that were invisible by other means. The true mathematician was accordingly the natural man, one who had retained a childlike connection to the natural world, unsullied by the corruptions of human society. Galois' story came into circulation at a time when notions of the nature of mathematics were shifting, and was emblematic of the novel idea that mathematics is its own alternative universe, accessible only to a select few who were gifted with special sight. Immersed as they are in this far-off land of truth and beauty, such geniuses are ill suited to life in our world and soon succumb to the machinations of petty, worldly men. Precisely the same is true of the iconic images of mathematicians in their contemporary portraits. The depictions of Johann Bernoulli and d'Alembert as prosperous worldly men were inseparable from the natural mathematics of the eighteenth century, just as the portraits of Galois and Abel as romantic dreamers were part and parcel of the new, otherworldly mathematics.

Finally, alongside the shifting stories and changing images, a profound transformation was taking place in the technical practice of mathematics. In the eighteenth century, as in previous ages, mathematics was taken to be the study of the world and its hidden harmonies. Even as analysis moved from the concrete physical problems that had occupied Johann Bernoulli at the beginning of the century to the highly abstract mathematical constructions of Lagrange at the century's close, the ultimate object of mathematical studies remained physical reality. Enlightenment mathematicians did not therefore much concern themselves with the finer points of mathematical rigor and consistency. If a mathematical construction adequately describes the world, they reasoned, then it must be essentially correct, and there was no need to worry too much about technical niceties. Reality itself, in other words, guarantees the correctness of mathematics.

All this, however, changed once mathematics came to be viewed as a self-contained world all its own. With no external reality to circumscribe it, mathematics was thrown back on its own resources to guarantee its truths. Shorn of its traditional roots in the physical world, mathematics could henceforth rely only on its own internal standards to determine the correctness of its results. As Cauchy, Galois, Abel, and Jacobi insisted, it was now essential to define the exact parameters of each mathematical question, to follow each argument with absolute precision, and never to allow physical intuition or looseness of reasoning to undermine the perfect rigor of an argument. If mathematics could not trust its own internal standards of rigor, then it could rely on nothing at all, and the entire edifice was but a house built on shifting sands.

Historians of mathematics have long noted the sharp transition from the intuitive mathematics of the eighteenth century to the modern rigorous mathematics invented in the early nineteenth century. Most often they have ascribed the shift to internal difficulties within mathematics, certain problems that could not be resolved in the old style and required a radically different approach. There is, no doubt, some truth in this perspective. Lagrange's investigations into the theory of equations, for example, had reached an impasse and could go no further without Galois' abstract (and seemingly useless) approach. Similarly, investigations into the parallel postulate had seemingly reached their limit in the late eighteenth century and could be carried forward only with the aid of radically new assumptions.

To a historian, however, there are dangers implicit in this view, which tends to assume that developments that did in fact take place needed necessarily to take place. But more crucially, this account fails to note that the profound shift in mathematical practice was inseparable from contemporaneous cultural developments. The move from the intuitive natural mathematics of the eighteenth century to the rigorous and self-contained mathematical style that appeared in the nineteenth century went hand in hand with a transformation in the stories told about mathematicians, their visual representations, and their self-fashioning. The geometer as a natural man, fully engaged with his physical and social surroundings, was replaced by the mathematician as a romantic hero, ill at ease in the world and sacrificing all for his higher calling. The image of d'Alembert as the paradigmatic mathematician was replaced by that of Galois.

But this is not all, for the cultural factors surrounding the shift in mathematical practices are not limited to a transformation in the mathematical persona. This transformation in the character and imagery of mathematics points the way toward the broader historical context in which mathematics was practiced. The image of the geometer as a natural man had its roots in Enlightenment notions of the purity of nature and the corruption of society, as well as in the age's persistent faith in the intelligibility of the physical world. It is hardly a coincidence that at the time of his death d'Alembert was hailed as a Rousseauian hero, an eternal child who, unlike most men, never lost his unmediated connection to nature. In the eyes of his contemporaries this made d'Alembert particularly well suited for the role of the geometer—uncovering the hidden harmonies that underlay the apparent copiousness of nature.

Similarly, the nineteenth-century image of the mathematician as a tragic hero is inseparable from the sensibilities of High Romanticism, a movement that was reaching its apex in the same years in which Galois, Abel, and Bolyai were making their mathematical discoveries. The young mathematical martyrs were depicted in ways that were indistinguishable from contemporaneous images of tragic poets, painters, and musicians. At that time the narratives associated with mathematics broke ranks with those associated with the natural sciences and aligned themselves instead with the stories and images associated with the creative arts. Just as the artists of the age were engaged in a pursuit of sublime experience that transcended the mundane reality of our world, so were the young mathematical poets pursuing an alternative reality of truth and beauty, completely separate from our imperfect universe.

It is clear from all this that the stories, myths, and images that accompany mathematical practices are far more than amusing anecdotes, serving to alleviate the stress of mathematical studies. They are, rather, essential accompaniments to mathematical work, defining the purpose and boundaries of mathematical practice, as well as pointing the way to the broader cultural context in which these practices flourished. In the eighteenth century the notion of the geometer as a natural man defined the outlines of mathematical practice by indicating that its subject matter was nature itself, and in the nineteenth century tales depicting mathematicians as romantic heroes suggested their object as being a different—and higher—reality. At the same time those stories were firmly grounded in the specific historical setting in which they originated—the Enlightenment cult of the natural man in the eighteenth century, and the romantic ideal of the tragic martyr in the nineteenth century. Mathematical stories, in other words, are deeply intertwined with both abstract mathematical work and its broader historical setting. As such, they serve to anchor mathematical practices in the cultural context that gave them birth.

The new mathematics practiced by Cauchy, Galois, and Abel did not take over the academic field overnight. In France, the established center of mathematical learning, mathematics was for decades to come dominated by graduates of the École polytechnique, who carried on the Enlightenment tradition of applied mathematics into the nineteenth century. In England, meanwhile, a different tradition emphasizing the geometrical foundations of mathematical practice took root and became institutionalized in the Cambridge University tripos exams. It was in Germany that the new mathematics found its true home. For a full century, beginning when the Prussian government tried belatedly to lure Abel to Berlin, Germany was the undisputed world leader in pure mathematical studies. A long line of brilliant German mathematicians, including Carl Gustav Jacobi (1804–1851), Pierre Gustav Lejeune Dirichlet (1805–1859), Karl Weierstrass (1815–1891), Leopold Kronecker (1823–1891), Bernhard Riemann (1826–1866), Richard Dedekind (1831–1916), Georg Cantor (1845–1918), Felix Klein (1849–1925), and David Hilbert (1862–1943), to name only the most illustrious, set the standard for the field and charted its direction. They took up the new rigorous mathematics where the tragic pioneers of the early nineteenth century left off, and they made it into a mathematical universe of a depth and complexity well beyond anything that Abel, Galois, and Cauchy had ever imagined. Through it all they congregated under the banner of Crelle's journal, whose quick rise to prominence and slow decline paralleled the fortunes of the German tradition of pure mathematics.

Only in the early decades of the twentieth century did the German ascendancy come to an end. The shock of World War I ruptured the international mathematical community, leaving the Germans isolated, and the wounds were slow to heal even once the war was over. The years of Nazi rule that followed ravaged German mathematics just as it did all German science, sending the best and the brightest fleeing to build their careers elsewhere. Meanwhile, competing centers of mathematical learning were gaining ground. In France a group of young mathematicians centered on the École normale broke away from the physicalist tradition of the École polytechnique. Publishing under the collective name Nicolas Bourbaki, they demanded a level of purity and rigor in mathematics past anything practiced by the German school. At the same time the United States was fast becoming a rival to the traditional European centers, thanks both to a dynamic homegrown tradition of pure mathematics and to an influx of European refugees. In the years after World War II, with the great cities of Europe lying in ruins, the United States became the world leader in mathematics, as in other scientific fields.

But through all the dramatic changes that reshaped the international mathematical community, the ideal of purity and rigor established in the early nineteenth century held firm. As the centers of pure mathematical research moved from France to Germany and then to America, the standards for mathematical argumentation were never relaxed but rather were tightened. Proofs that seemed perfectly adequate to Cauchy appeared sloppy to Riemann and Weierstrass, whose own arguments, in turn, seemed naïvely lacking in rigor to subsequent generations. This trend was probably inevitable, since as long as mathematics was understood to be its own self-contained world, divorced from physical reality, there was no escape from an ever-increasing insistence on the only standard of truth left to the field—strict internal rigor. And as this new, self-supporting mathematics survived and flourished, so did its accompanying myth of the tragic mathematician.

The legends of Abel, Bolyai, and especially Galois became founding myths of mathematics, repeated from one generation of practitioners to the next. They were, and still are, emblematic tales, elucidating what true mathematics and a true mathematician should be. Furthermore, as we have seen, the paradigm they established of the "tragic mathematician" survived not only in the tales of these past giants of the field but also in the lives of some of the greatest practitioners in each succeeding generation. There was Riemann, the most brilliant mathematician of the second half of the nineteenth century, who died at 40 and whose papers were set ablaze by an overeager housekeeper. There was Cantor, driven mad by his obsession with transfinite numbers and lack of professional recognition. Srinivasa Ramanujan, the tragic Indian genius, was another, as was (to an extent) his mentor, G. H. Hardy. The list goes on: Kurt Gödel, Alan Turing, John Nash, and Alexander Grothendieck. Most recent is the case of Grigory Perelman, a present-day eccentric genius who proved the Poincaré conjecture but who (according to reports) abandoned mathematics when a rival tried to lay claim to his discovery.

How it came about in each case that so many of the greatest mathematicians of the past two centuries lived tragically romantic lives reminiscent of those of Galois and Abel matters little. It may have been that a mathematician's own self-understanding motivated him—consciously or otherwise—to fashion a life modeled on the example of the tragic heroes of the early nineteenth century. Or it may have been that a mathematician's contemporaries molded his story, however unsuitable it may have been, into a familiar shape that suited their own expectations. Galois himself, who saw himself as a persecuted genius, is an example of the former; Abel, a conventional and sociable young man who was canonized by Crelle, Libri, and others after his untimely death, is an example of the latter. But in either case the result was the same: the canonical stories were preserved in the collective memory of the mathematical community and lived again and again throughout the generations.

For the past two centuries the study of pure mathematics has been inseparable from the story of the mathematician as a tragic romantic hero. The notion that mathematics exists as a self-contained alternative reality and the conviction that all too often the purest mathematical genius is misunderstood by his contemporaries and doomed to persecution by an uncaring world have gone hand in hand since the birth of the new mathematics in the early nineteenth century. Indeed, the human story and the mathematical practice are two sides of the same coin, each complementing and supporting the other. If mathematics indeed resides on a different and higher plane of reality, would not its practitioners be those rare geniuses unaccountably gifted with the ability to see into this alternative universe? Furthermore, since mathematical reality is completely dissociated from our familiar world, it stands to reason that these pure souls would be ill suited for daily life and be misunderstood by less gifted men. Conversely, if mathematicians are indeed, as the story suggests, isolated and persecuted geniuses, what is that study that has so completely captivated them as to make them unsuitable for life in our world? It must be a reality so strange and so rich that it absorbs them, mind and soul, a world all its own, inaccessible to ordinary mortals. In fact, the more alienated our mathematical geniuses are from their surroundings, and the more they are persecuted by lesser men, the stronger the proof of just how different the mathematical world is from our mundane existence. In this manner the new mathematical practice suggests the romantic story, while the story, in turn, points to a new, otherworldly mathematics.

This interdependence between the story and the practice, the mathematical poet and the new rigorous mathematics, does not in itself tell us whether one preceded or instigated the other. Did the new romantic sensibility of the early nineteenth century, with its accompanying legends, bring about a radically new approach to mathematics? Or, conversely, did a new mathematics that appeared on its own require a new founding myth, which it happened to find in the tragic legends of the romantic poets and artists? Neither of these options alone seems probable. Mathematics is an ancient and enduring intellectual tradition, and a "new mathematics" does not spring to life simply because romantic poets would wish it to. At the same time it is undeniable that early nineteenth-century mathematicians did not just build on the accomplishments of their predecessors but broke away from their practices in ways that their elders would have found abhorrent. That this development occurred for purely internal mathematical reasons seems implausible. That it is unrelated to the romantic movement, and to the remarkable romantic personas the new mathematicians took on at the very same time, stretches credulity.

Perhaps the closest one can get to an elucidation of the precise relationship between the new rigorous mathematical practice and the new romantic mathematical stories is this: the new romantic sensibilities of the early nineteenth century allowed mathematics to develop in certain directions that were previously considered illegitimate, opening new venues that made the work of Galois, Abel, Cauchy, and their successors possible. In the eighteenth century geometers such as d'Alembert and Lagrange always took care to keep their mathematics, even at its most abstract, as close as possible to the physical world. Failing to do so would open them to the charge—made by Diderot and others—that mathematics was a clever but pointless mind game. But the romantics' idealization of the pursuit of the sublime, of the search for truth and beauty that cannot be found in mundane experience, opened up other possibilities. By adopting the romantic outlook, and with it the romantics' tragic persona, mathematicians were free to develop their field in a new direction. It was no longer necessary to keep their studies bound to physical reality; in fact, it was much preferable to keep mathematics as far as possible from worldly contamination. Mathematics could now legitimately pursue higher truths all its own, unsullied by physical reality and freed of the charge of irrelevance. After all, in the romantics' view, nothing could be more relevant and worthy than the pursuit of truth for its own sake.

Such questions about the interrelationship between the romantic milieu and the development of a specific field are not, of course, unique to mathematics. In what ways was the poetry of Byron and Keats shaped by the romantic movement? Or the paintings of Delacroix, or the music of Beethoven? In each case their work is clearly imbued with new motifs and sensibilities while at the same time building on centuries-long traditions of poetry, painting, or music. Precisely elucidating the interplay between the ancient tradition and the new romantic beginning is never easy and perhaps impossible. But no one would deny Keats, Delacroix, Beethoven, and their numerous contemporaries a place in the broad cultural movement known as Romanticism. Much the same is true of the new mathematicians of the romantic age. Pointing out the exact interplay between the long-standing mathematical tradition and the break initiated by Romanticism is difficult, but this should not deny the new mathematicians a place in the broader cultural movement of their time. In their lives and in their works, Galois, Abel, Cauchy, and Bolyai were as much figures of the romantic age as Byron, Shelley, Delacroix, and Beethoven.

Nevertheless, although the new mathematics had its roots in the age of Romanticism, there is no denying that it has lasted much longer than the romantic movement itself. Not only have the tragic lives of iconic mathematicians repeated the pattern set by the mathematical poets of the early nineteenth century, but the fundamental assumptions that govern the practice of mathematics have also persisted from their day to ours. The notion that true mathematics is its own self-contained world, separate from physical reality, ruled by pure reason, and safeguarded by strict standards of rigor, seemed radical when it was first proposed by Cauchy, Abel, and their contemporaries. But nearly two centuries later these tenets appear so obvious to professional mathematicians that they are hardly ever stated. The articles, books, lectures, and courses written and presented by modern academic mathematicians all treat these assumptions as self-evident and in no need of elaboration. In 1940 G. H. Hardy in *A Mathematician's Apology* provided a beautifully written and well-articulated defense of this attitude, emphasizing that it is shared by most practicing mathematicians. Three-quarters of a century later few mathematicians have felt the need to add to Hardy's exposition, and his essay remains the most definitive account of the craft of a mathematician addressed to the general public.[8]

Are we therefore to assume that the mathematical practice and the mathematical stories first introduced in the age of High Romanticism are here to stay? Has mathematics, in other words, reached its final form? Probably not. The persistence of the romantic style of mathematics in a fast-changing world certainly points to the strength of the internal tradition passed down from one generation of mathematicians to the next. It demonstrates the remarkable fruitfulness of this mathematical practice, as well as the social coherence of the international community of pure mathematicians. Nevertheless, mathematics was and remains an intellectual activity grounded in cultural and historical circumstances. And just as the practice of mathematics was transformed in the past, so it can change again.

One possible direction of change is indicated by the current debate on the use of computers in mathematical proofs. Through their sheer computing power, unimaginable to previous generations of mathematicians, computers can test the validity of a theorem in a seemingly infinite number of cases. They can, in other words, "prove" a theorem by brute force, showing it to be either true or false not by elegant abstract reasoning but by going over all—or virtually all—possibilities case by case. Such an experimental approach to mathematical proof is not new. English mathematician John Wallis, for example, one of the leading mathematicians of the seventeenth century, openly advocated and practiced a mathematical style in which a statement was checked for a few cases and then generalized. Furthermore, to some extent such experimentalism has been practiced by all mathematicians in their research, even if they did not consider it a finished product without the formal deductive proof. But no human can approach a computer's capacity to test and confirm the validity of a theorem and, perhaps, prove it for all practical purposes.[9]

Inevitably the very notion of proof by computer is a monstrosity to most academic mathematicians, adherents of pure mathematics as practiced since the days of Cauchy. A proof "for all practical purposes" may have been perfectly satisfactory for Enlightenment mathematicians such as d'Alembert and Euler, but it is unacceptable to mathematicians schooled in the rigorous mathematics of the past two centuries. Such a proof is by definition nonrigorous and risks the grave errors of reasoning that Abel had warned of. To make things worse, proofs by computer are not beautiful or elegant in the traditional mathematical sense but rely on brute computational force and the ability to try out huge numbers of individual cases. This, to traditional mathematicians, is a fatal flaw: as G. H. Hardy wrote, "Beauty is the first test: there is no place in the world for ugly mathematics."[10]

Finally, and perhaps worst of all, proofs by computer compromise the purity of mathematics by engaging it too closely with the material world. The whole point of the new mathematics forged in the early nineteenth century was to free the field once and for all from its dependence on physical reality. Instead of a field defined as an abstraction from the physical world, the new mathematics was a self-contained world governed solely by mathematical—not physical—principles. But introducing computers into mathematical practice would immediately change all that. Computers, after all, are machines, made of silicon and wires, chips, motherboards, and liquid crystal displays. They are stubbornly and irrevocably material, unapologetically a part of the physical world. Allowing such instruments into the practice of higher mathematics would contaminate the field and dilute its purity. It would undermine pure mathematics' hard-won independence, making it once more dependent on physical reality. To many academic mathematicians, this would be a betrayal of all that mathematics is and should be.

The stakes in the debate over the role of computers in higher mathematics are therefore high. Legitimizing the use of computers in mathematical proofs would profoundly change the nature of the field and take it in a direction very different from the one it has been following for the past two centuries. The advocates of computers are currently a small, if dynamic, minority, and most pure mathematicians tend to dismiss the very notion of proof by computer. With such strong opposition it is still too early to predict whether the new approach will ultimately gain legitimacy, or whether the traditional practice of pure mathematics will weather the controversy and continue on its current course. Nevertheless, it may not be too early to speculate on whether a new mathematical practice in which computers take a major role will be accompanied by a new and different narrative—a new legend, if you will, of the life and role of a mathematician.

Notably, the popular image of a computer expert or "wiz" is very different from that of a mathematical genius. Like the iconic mathematician, the computer wiz spends his life in a universe of signs distinct from our familiar world, a reality made of binary symbols and computer languages indecipherable to ordinary humans. Like the mathematical genius, furthermore, the computer wiz is something of a social outcast. He (for the type is unequivocally male) is a "nerd," so completely engaged in his alternative reality that he is incapable of normal human relations, has no social life, and spends his time in his proverbial "parents' basement."

Here, however, the similarities end, for there is nothing tragic or heroic, not to mention romantic, about the figure of the computer nerd. The ideal mathematical genius, as we have seen, is a helpless innocent, so irresistibly drawn to the higher realms that are his true home that he is incapable of dealing with the mundane realities of daily life. The computer wiz also dwells in an alternative world of signs, but he keeps his gaze focused firmly on our own material world, nursing all the injuries done to him by fellow humans and secretly plotting his revenge. He is never in danger of losing himself in a pure alternative reality, because the world of computer signs is not a higher reality but an artificial one: he is its creator and its master. His knowledge of the arcane mysteries of computers is for him a means to worldly ends and is put in the service of an unquenchable thirst for power. Not for nothing is he known as a wizard, for he is indeed the heir of the Renaissance magus, a manipulator of arcane signs to gain power over earthly reality. The guiding story of the traditional mathematical genius was the tragic tale of Galois, a young martyr to mathematical purity; the popular myth of the computer genius sounds more like *The Revenge of the Nerds*.

It is too early to say whether proof by computer will ultimately gain mainstream acceptance among academic mathematicians. But if it does, it is not unlikely that the narrative of the power-hungry nerd will encroach, to some extent, on the hagiographic accounts of mathematical martyrs. Just as the silicon chips of a computer could taint the ethereal purity of higher mathematics, so the unsavory computer wiz could taint the pristine image of the mathematical saint. Quite likely the end result will be a new mathematical persona altogether, incorporating elements from both narratives in a way that is fundamentally different from either. It is also quite possible that the challenge to the current prevailing mathematical myth will come from an altogether different direction that is not even imaginable at present. There is no way to tell. The only certainty is that mathematics will change once again, as it has in the past. And when it does, it will be accompanied, as it always has, by a new compendium of stories.

NOTES

INTRODUCTION

1. This does not, of course, mean that mathematicians could not suffer misfortune. Johann Samuel König, a mathematician who dared to challenge the great Euler and Pierre-Louis Moreau de Maupertuis, president of the Berlin Academy, was effectively banished from the Republic of Letters and died shortly afterward. But unlike his successors in the nineteenth century, König's suffering did not earn him a reputation as a genius or a martyr for mathematics. His banishment was a misfortune, not a martyrdom, and he was soon forgotten.

2. Whereas the story of Hippasus points to the dangers of unrestricted mathematical reasoning, the story of Aristippus suggests the opposite view, that geometry is a safe harbor of certainty within the raging storm of intellectual disputes. For more on the story of Aristippus and its significance, see Amir R. Alexander, *Geometrical Landscapes: The Voyages of Discovery and the Transformation of Mathematical Practice* (Stanford: Stanford University Press, 2002), 198–200.

3. On the exploration narrative and its transformation of mathematics, see Alexander, *Geometrical Landscapes;* Amir R. Alexander, "Exploration Mathematics: The Rhetoric of Discovery and the Rise of Infinitesimal Methods," *Configurations* 9 (Winter 2001): 1–36; Amir R. Alexander, "The Imperialist Space of Elizabethan Mathematics," *Studies in the History and Philosophy of Science* 26, no. 4 (December 1995): 559–591; and Mary Terrall, "Mathematics in Narratives of Geodetic Expeditions," *Isis* 97, no. 4 (December 2006): 683–699.

1. THE ETERNAL CHILD

1. Thomas L. Hankins, *Jean d'Alembert: Science and the Enlightenment* (New York: Gordon and Breach, 1990; first published, Oxford: Oxford University Press, 1970), 18–19.

2. Ibid., 18.

3. On d'Alembert's patronage of Condorcet and their association over several decades, see Keith Michael Baker, *Condorcet: From Natural Philosophy to Social Mathematics* (Chicago: University of Chicago Press, 1975).

4. Condorcet, "Éloge de M. d'Alembert," in *Oeuvres complètes de Condorcet,* vol. 3 (Brunswick: Vieweg, 1804; Paris: Heinrichs, 1804), 76–160.

5. Ibid., 76. Translation here and elsewhere (unless indicated otherwise) is by the author.

6. On d'Alembert's and Condorcet's views on the role of philosophes, see Baker, *Condorcet,* 12–16, 193–194, and passim.

7. Condorcet, "Éloge de M. d'Alembert," 76–77.

8. Ibid.

9. Hankins, *Jean d'Alembert,* 19.

10. Condorcet, "Éloge de M. d'Alembert," 82–83.

11. Ibid., 83.

12. Ibid., 80.

13. Ibid.

14. Ibid., 81.

15. Ibid.

16. Thomas Hobbes, *Leviathan* (Oxford: Oxford University Press, 2009; first published in 1651); Locke, *Second Treatise of Government* (Indianapolis: Hackett Publishing Company, 1980; first published in 1690); Jean-Jacques Rousseau, *The Social Contract* (London: Penguin Books, 1968; first published as *Du contrat social: Principes du droit politique* [Paris, 1762]).

17. Jean-Jacques Rousseau, *Emile; or, On Education* (New York: Basic Books, 1979; first published as *Émile, ou l'Éducation* [Paris, 1762]). The quote is from 214, early in book 4.

18. Ibid., book 4, 327–333.

19. Condorcet, "Éloge de M. d'Alembert," 81.

20. Ibid., 80.

21. Ibid., 83.

22. Ibid.

23. Alexis Clairaut, quoted in Hankins, *Jean d'Alembert,* 31.

24. On the rivalry between d'Alembert and Euler, see ibid., 42–62.

25. Jean d'Alembert, *Réflexions sur la cause générale des vents* (Paris: David, 1747).

26. Hankins, *Jean d'Alembert,* 47–48.

27. Jean d'Alembert, *Recherches sur la précession des équinoxes et sur la nutation de l'axe de la terre dans le systême newtonien* (Paris: David, 1749); d'Alembert, *Essai d'une nouvelle théorie de la résistence des fluids* (Paris: David, 1752).

28. Frederick II of Prussia, quoted in Florian Cajori, "Frederick the Great on Mathematics and Mathematicians," *American Mathematical Monthly* 34, no. 3 (March 1927): 122.

29. Ibid., 123.

30. Ibid., 126. For more on Frederick's correspondence with Maupertuis, see Mary Terrall, *The Man Who Flattened the Earth: Maupertuis and the Sciences in the Enlightenment* (Chicago: University of Chicago Press, 2002), 181.

31. Cajori, "Frederick the Great," 125.

32. Frederick II of Prussia, quoted in Hankins, *Jean d'Alembert,* 62.

33. See ibid., 56n3.

34. Ibid., 60.

35. Quoted in Cajori, "Frederick the Great," 128.

36. Frederick II of Prussia, quoted in ibid., 122.

37. On Maupertuis, see Terrall, *Man Who Flattened the Earth;* on the Lapland expedition, see especially chapter 4, and on Frederick's courting of Maupertuis, see especially chapter 6. See also Mary Terrall, "Representing the Earth's Shape: The Polemics Surrounding Maupertuis's Expedition to Lapland," *Isis* 83 (1992): 218–237.

38. Friedrich Melchior von Grimm, *Correspondence littéraire, philosophique, et critique,* ed. Maurice Tourneux, 16 vols. (Paris, 1877–1882), 13:461; quoted in Hankins, *Jean d'Alembert,* 16–17.

39. Johann Lexell, quoted in Hankins, *Jean d'Alembert,* 17. See Arthur Birembaut, "L'Académie royale des sciences en 1780 vue par l'astronome suédois Lexell (1740–84)," *Revue d'Histoire des Sciences* 10 (April–June 1957): 148–166.

40. Condorcet, "Éloge de M. d'Alembert," 96.

41. Hankins, *Jean d'Alembert,* 16.

42. Jean le Rond d'Alembert, *Preliminary Discourse to the Encyclopedia of Diderot*, trans. Richard N. Schwab (Chicago: University of Chicago Press, 1995; first published as "Discours Preliminaire de l'Encyclopédie" [Paris, 1751]).

43. On d'Alembert's efforts to shape the composition of the two French academies with the backing of salon society, see Baker, *Condorcet,* 16–23; and Hankins, *Jean d'Alembert,* chap. 6.

44. On the philosophical rift between d'Alembert and Diderot, see Hankins, *Jean d'Alembert,* chap. 4; and Denis Diderot, "D'Alembert's Dream," in Denis Diderot, *Rameau's Nephew and d'Alembert's Dream*, trans. Leonard Tancock (London: Penguin Classics, 1978).

45. Hankins, *Jean d'Alembert,* chaps. 4, 6.

46. Condorcet, "Éloge de M. d'Alembert," 154.

47. Ibid.

48. Ibid., 81.

49. Ibid., 80.

50. Ibid., 95–96.

51. Ibid., 96.

52. Maupertuis, for one, believed that his "principle of least action" proved God's providence. Euler was a devout minister's son, and d'Alembert was a deist. Laplace, according to legend, when asked by Napoleon where God fit into his world system, responded, "I have no need of that hypothesis." On Maupertuis, see Terrall, *Man Who Flattened the Earth;* on Euler, see Adolph P. Youschkevitch, "Euler, Leonhard," in *Dictionary of Scientific Biography*, ed. Charles P. Gillispie, 16 vols. (New York: Scribner, 1970–1980); on d'Alembert, see Hankins, *D'Alembert;* on Laplace, see Eric T. Bell, *Men of Mathematics* (New York: Simon and Schuster, 1986; first published 1937), 181.

53. Bernard Le Bovier de Fontenelle, "Éloge de Monsieur Varignon," in Fontenelle, *Éloges des académiciens avec l'histoire de l'Académie royale des sciences en MDCXCIX* (The Hague: Isaac vander Kloot, 1740), 2:175–194. The quote is from page 192. Translation of the Eulogies, here and elsewhere, is by the author.

54. Condorcet, "Éloge de M. d'Alembert"; Condorcet, "Éloge de Euler," in *Oeuvres complètes de Condorcet,* 3:3–62.

55. Condorcet, "Éloge de M. Euler," 46.

56. Ibid., 56.

57. Ibid., 62.

58. Denis Diderot, "Thoughts on the Interpretation of Nature" [1753], in Denis Diderot, *Thoughts on the Interpretation of Nature and Other Philosophical Works,* ed. David Adams (Manchester, UK: Clinamen Press, 1999), 59.

59. Diderot, "D'Alembert's Dream." Diderot was not alone in viewing mathematics as disconnected from the world. See also the Comte de Buffon, "Premier discours," in *From Natural History to the History of Nature: Readings from Buffon and His Critics,* ed. and trans. John Lyon and Philip R. Sloan (Notre Dame, IN: University of Notre Dame Press, 1981). For more on Enlightenment mathematics and its philosophical roots, see Chapter 2.

2. NATURAL MATHEMATICS

1. D'Alembert, *Preliminary Discourse,* 20–21.

2. Ibid., 20.

3. Ibid., 21.

4. This refers to René Descartes' famous dictum "I think, therefore I am" ("Cogito, ergo sum"), which he uses to establish his own existence in the face of all-encompassing doubt.

5. D'Alembert, *Preliminary Discourse,* 25.

6. Buffon, "Premier discours."

7. Elsewhere Diderot wrote: "The domain of mathematicians is a world purely of the intellect, where what are taken for absolute truths cease entirely to be so when applied to the world we live in." See Diderot, "Thoughts on the Interpretation of Nature," 35.

8. Denis Diderot, "Letter on the Blind" [1749], in Denis Diderot, *Thoughts on the Interpretation of Nature and Other Philosophical Works,* ed. David Adams (Manchester, UK: Clinamen Press, 1999), 147–200.

9. On d'Alembert and eighteenth-century "natural" mathematics, see Joan L. Richards, "Historical Mathematics in the French Eighteenth Century," *Isis* 97, no. 4 (2006): 700–713.

10. According to historian of mathematics Craig Fraser, for example, "In the 18th century . . . an *a-priori* belief in the intrinsic mathematical character of physical reality continued to underlie work in mechanics." He continues: "Mechanics was regarded either as a development of geometry obtained by the introduction of the concepts of impenetrability and time, or as something that, like geometry, was reducible to the study of relations among analytical variables." See Craig Fraser, "Lagrange's Analytical Mathematics, Its Cartesian Origins and Reception in Comte's Positive Philosophy," *Studies in History and Philosophy of Science* 21, no. 2 (1990): 243–256; quotes on 245 and 250–251, respectively.

11. On the character of eighteenth-century mathematics and the movement from the geometrically concrete to the algebraically general and abstract, see Thomas L. Hankins, *Science and the Enlightenment* (Cambridge: Cambridge University Press,

1985), chap. 2; Hans Niels Jahnke, "Algebraic Analysis in the 18th Century," in Hans Niels Jahnke, ed., *A History of Analysis* (Providence, RI: American Mathematical Society and London Mathematical Society, 2003; first published in German in 1999), 105–108; and H. J. M. Bos, "Differentials, Higher Order Differentials, and the Derivative in the Leibnizian Calculus," *Archive for History of Exact Sciences* 14 (1974): 1–90, esp. 1–12.

12. A somewhat different mathematical tradition prevailed in Britain, where throughout the eighteenth and nineteenth centuries analysis remained far closer to its geometrical roots than it did on the Continent. See Joan L. Richards, *Mathematical Visions: The Pursuit of Geometry in Victorian England* (Boston: Academic Press, 1988).

13. Leonhard Euler, quoted in Jahnke, "Algebraic Analysis in the 18th Century," 107.

14. Joseph-Louis Lagrange, "Preface to the First Edition," in *Analytical Mechanics*, trans. and ed. Auguste Boissonnade and Victor N. Vagliente, Boston Studies in the Philosophy of Science, vol. 191 (Dordrecht: Kluwer Academic Publishers, 1997), 7. This is a translation of the new edition of 1811, which includes the preface to the first edition of 1788.

15. On Euler's and Lagrange's views on algebraic analysis, including the values of abstraction and generalization, and its usefulness in describing the physical world, see Jahnke, "Algebraic Analysis in the 18th Century," 106–136; Craig G. Fraser, "Mathematics," in *The Cambridge History of Science*, vol. 4 (Cambridge: Cambridge University Press, 2003), 305–327; Fraser, "Lagrange's Analytical Mathematics," 243–256; Craig G. Fraser, "The Calculus as Algebraic Analysis," *Archive for History of Exact Sciences* 39 (1989): 317–335; and Craig G. Fraser, "Joseph-Louis Lagrange's Algebraic Vision of the Calculus," *Historia Mathematica* 14 (1987): 38–53.

16. D'Alembert, *Preliminary Discourse*, 20–21.

17. This suggestion is voiced by Salviati toward the end of the fourth day of discussions in Galileo Galilei, *Dialogues Concerning the Two New Sciences*, trans. Henry Crew and Alfonso de Salvio (Buffalo, NY: Prometheus Books, 1991; first published, New York: Macmillan, 1914), 290. This is a translation of Galileo's last book, the *Discorsi e demonstrazioni mathematiche* of 1638.

18. Johann Bernoulli's analysis of the catenary can be found in in his *Lectiones mathematicae de methodo integralium* (Paris, 1692), republished in *Johannis Bernoulli opera omnia*, vol. 3 (Lausanne: Bousquet, 1742), 385–558. The discussion of the catenary can be found on pages 491–497. For a discussion of Bernoulli's treatment of the catenary, see Jahnke, "Algebraic Analysis in the 18th Century," 109–110.

19. For a nuanced discussion of the practices of the Leibnizian calculus, see Bos, "Differentials." Bos emphasizes that the Leibnizian calculus dealt with fixed relations between the differentials dx and dy, which were on an equal footing with each other. This is different from the modern calculus, in which the dependent variable y is a function of the independent variable x, which moves through a range of values. In the modern designation, $dy{:}dx$ is not the "differential" of a curve but the "derivative" of the function $y = f(x)$.

20. Note that Bernoulli's practice here is the opposite of the modern convention in which the horizontal axis is designated *x* and the vertical axis *y*. That would make *Ha* the differential *dx* and *AH* the differential *dy*.

21. Bernoulli, *Johannis Bernoulli opera omnia,* 3:492.

22. For a discussion of this problem, see Hankins, *Science and the Enlightenment,* 22–25.

23. For a discussion of the polygonal curve with an infinite number of sides as an approximation of a "smooth" curve in Leibniz's calculus, see Bos, "Differentials," esp. 15–19. These passages also discuss the tangent to a curve as a continuation of the side of a polygon with an infinite number of sides.

24. On Louville, see Hankins, *Science and the Enlightenment,* 24.

25. On the vibrating-string debate, see Jahnke, "Algebraic Analysis in the 18th Century," 123–127; and Hankins, *Jean d'Alembert,* 47–49.

26. Daniel Bernoulli's contribution to the vibrating-string debate was contained in an article in the *Mémoires* of the Berlin Academy. See Jahnke, "Algebraic Analysis in the 18th Century," 125.

27. Euler explained his views in his *Elements of Algebra* of 1771, published in English translation in 1972; quoted in Jahnke, "Algebraic Analysis in the 18th Century," 107.

28. On Enlightenment mathematicians' preference for algebra from the mid-eighteenth century onward, see Jahnke, "Algebraic Analysis in the 18th Century"; Fraser, "Calculus as Algebraic Analysis"; Fraser, "Mathematics"; Fraser, "Lagrange's Analytical Mathematics"; Fraser, "Joseph-Louis Lagrange's Algebraic Vision of the Calculus"; and Bos, "Differentials."

29. On Newtonians and Cartesians in early eighteenth-century France, see Mary Terrall, *Man Who Flattened the Earth;* Hankins, *Science in the Enlightenment*, chap. 3; Hankins, *Jean d'Alembert*, chap. 1; and J. B. Shank, *The Newton Wars and the Beginning of the French Enlightenment* (Chicago: University of Chicago Press, 2008), passim.

30. Jean d'Alembert, *Traité de dynamique,* reprint of the second edition, with a new introduction and bibliography by Thomas L. Hankins (New York: Johnson Reprint Corporation, 1968; first published, Paris: David, 1758), v–vi. Also cited in the same volume, introduction, xviii.

31. On Cartesian strains in d'Alembert's thought, see Hankins, *Jean d'Alembert;* and Hankins, introduction to d'Alembert, *Traité de dynamique*, xviii–xxiii.

32. Jean d'Alembert, quoted in Hankins, introduction to d'Alembert, *Traité de dynamique,* xxix. D'Alembert's principle states (in modern terms) that the impressed forces on a mechanical system are balanced by the sum of the inertial forces and the forces of constraint. See Hankins, introduction, xxxii.

33. Leonhard Euler, "De differentatione functionum duas pluresve variabiles quantitates involventum," manuscript F.136, op. 1, no. 97, 2–9, in the archives of the Russian Academy of Sciences in St. Petersburg. The text of the manuscript and its translation into English can be found in Steven B. Engelsman, *Families of Curves and the Origins of Partial Differentiation* (Amsterdam: Elsevier Science Publishers, 1984), appendix 2, 204–222. The treatise is also discussed in Engelsman, *Families*

of Curves, 124–132. Euler's proof of the equality of mixed partial differentials is also discussed in Fraser, "Calculus as Algebraic Analysis," 317–335, esp. 319–320.

34. On Nicolaus Bernoulli's work on the topic and its influence on Euler, see Engelsman, *Families of Curves,* chaps. 4, 5.

35. In the original Euler does not make use of the functions *S(x,a)* and *T(x,a),* but moves directly from the diagram to the final result. This makes his argument even more geometrical than it appears here. In the interest of clarity I follow Engelsman in making use of these auxiliary functions. See the text of "De differentiatione" in Engelsman, *Families of Curves,* appendix 2.

36. Engelsman, *Families of Curves,* appendix 2, 215.

37. In "De differentiatione" Euler in fact does not speak of "functions" but of "quantities." Later in his career, however, Euler was largely responsible for the development and dissemination of the concept of function, an issue that was at the center of his controversy with d'Alembert about the vibrating string. In the interest of clarity I use the concept of function here, a few years before Euler actually made use of it.

38. Engelsman, *Families of Curves,* appendix 2, 215–216. The algebraic proof was eventually published by Euler in 1735 in the treatise "De infinitis curvis eiusdem generis seu methodus inveniendi aequationes pro infinitis curvis eiusdem generis," which appears in Leonhard Euler, *Opera Omnia,* ser. 1, vol. 22 (Basel: Societatis Scientiarum Naturalium Helveticae, 1936), 36–75.

39. Lagrange, "Preface to the First Edition," 7.

40. Joseph-Louis Lagrange to Jean Lerond d'Alembert, 21 September 1781; quoted in Hankins, *Jean d'Alembert,* 99–100.

41. Lagrange, "Preface to the First Edition," 7.

42. D'Alembert, *Preliminary Discourse,* 20.

3. A HABIT OF INSULT

1. E. T. Bell, *Men of Mathematics* (New York: Simon and Schuster, 1986; first published in 1937).

2. Freeman Dyson, *Disturbing the Universe* (New York: Harper and Row, 1979), 14, quoted in Tony Rothman, "Genius and the Biographers," *American Mathematical Monthly* 89, no. 2 (1982): 84.

3. Newman, Hoyle, and Infeld are all quoted in Rothman, "Genius and the Biographers," 84. The movie is mentioned in René Taton, "Évariste Galois and His Contemporaries," *Bulletin of the London Mathematical Society* 15, pt. 2, no. 53 (1983): 107n2.

4. Laura Toti Rigatelli, *Évariste Galois (1811–1832)* (Basel: Birckhauser, 1996); Mario Livio, *The Equation That Couldn't Be Solved: How Mathematical Genius Discovered the Language of Symmetry* (New York: Simon and Schuster, 2005); Tom Petsinis, *The French Mathematician: A Novel* (New York: Walker and Company, 1998).

5. Jimmy Tseng, "Évariste Galois," in *ISCID Encyclopedia of Science and Philosophy,* http://www.iscid.org/encyclopedia/Evariste_Galois, accessed June 28, 2009.

6. Rothman, "Genius and the Biographers," 87–89; Livio, *Equation That Couldn't Be Solved,* 132–133.

7. Livio, *Equation That Couldn't Be Solved,* 133.

8. George Sarton, "Évariste Galois," *Osiris,* vol. 3 (1937): 246. This article was first published in *Scientific Monthly,* October 1921, 363–375. All citations are from the *Osiris* publication.

9. Livio, *Equation That Couldn't Be Solved,* 126–127. Here and subsequently, Livio and Rothman base their accounts of Galois' troubles on Paul Dupuy, "La vie d'Evariste Galois," *Annales scientifiques de l'École normale supérieure* 13 (1896): 197–266, 220–233.

10. Sophie Germain, quoted in Rothman, "Genius and the Biographers," 91. Cauchy, an ultraconservative royalist, had refused to sign an oath of loyalty to Louis-Philippe and had subsequently embarked on an eight-year self-imposed exile. René Taton puts the date of Libri's lecture as September 20, 1830, when Libri gave a presentation on the solution of a class of algebraic equations. See Taton, "Évariste Galois and His Contemporaries," 107–118, 112.

11. Livio, *Equation That Couldn't Be Solved,* 130–131.

12. Ibid., 131–132; Rothman, "Genius and the Biographers," 92–93.

13. Livio, *Equation That Couldn't Be Solved,* 135; Rothman, "Genius and the Biographers," 93–94; Dupuy, "Vie d'Evariste Galois," 238.

14. "Philippe portera sa tête sur ton autel, ô liberté!" Dupuy, "Vie d'Evariste Galois," 238.

15. Livio, *Equation That Couldn't Be Solved,* 137; Rothman, "Genius and the Biographers," 94ff.; Dupuy, "Vie d'Evariste Galois," 240–243.

16. François-Vincent Raspail, *Réforme pénitentiaire: Lettres sur les prisons de Paris,* vol. 2 (Paris: Tamisey et champion, 1839), 84; quoted in Rothman, "Genius and the Biographers," 94.

17. Raspail, *Lettres sur les prisons de Paris,* quoted in Rothman, "Genius and the Biographers," 94–95; Livio, *Equation That Couldn't Be Solved,* 136–139; Dupuy, "Vie d'Evariste Galois," 240–243.

18. Leopold Infeld, *Whom the Gods Love: The Story of Evariste Galois* (New York: Whittlesey House, 1948), 308–311.

19. Fred Hoyle, *Ten Faces of the Universe* (San Francisco: W. H. Freeman, 1977), chap. 1.

20. Toti Rigatelli, *Evariste Galois.*

21. Rothman, "Genius and the Biographers," 99; Livio, *Equation That Couldn't Be Solved*, 152.

22. Carlos Alberto Infantozzi discovered the identity of Galois' object of desire—Stéphanie Poterin du Motel. See Carlos A. Infantozzi, "Sur la morte d'Évariste Galois," *Revue d'Histoire des Sciences* 21, no. 2 (1968): 157–160.

23. Livio, *Equation That Couldn't Be Solved,* 141–142.

24. Ibid., 143.

25. Ibid., 144.

26. Ibid., 147.

27. Quoted ibid., 132–133.

28. Quoted ibid., 133.

29. Ibid., 132.

30. "Spiteful critics" is a loose translation of the original "zoïles."

31. Niels Henrik Abel was a Norwegian mathematical genius who died in 1829 at the age of 26 after allegedly being ignored by the Paris mathematical establishment. See Chapter 4 for a detailed discussion of Abel.

32. Évariste Galois, "Préface," in Robert Bourgne and J.-P. Azra, *Écrits et mémoires mathématiques d'Évariste Galois* (Paris: Gauthier Villars, 1962), 3–5. Translation based mostly on Rothman, "Genius and the Biographers," 97.

33. Évariste Galois, "Préface," in Bourgne and Azra, *Écrits et mémoires mathématiques d'Évariste Galois,* 7.

34. Évariste Galois to Auguste Chevalier, May 25, 1832, in Bourgne and Azra, *Écrits et mémoires mathématiques d'Évariste Galois,* 469.

35. "Évariste Galois," *Magasin Pittoresque* 16 (1848): 227–228. The article is unsigned, but Dupuy attributes it to Galois' former classmate, Pierre-Paul Flaugergues. See Dupuy, "Vie d'Evariste Galois," 250.

36. Auguste Chevalier, "Évariste Galois," *Revue Encyclopédique,* September 1832, 744.

37. Ibid., 746–747.

38. Ibid., 747.

39. Ibid., 749.

40. Ibid., 750.

41. Ibid., 751.

42. Ibid., 754.

43. Raspail, *Réforme pénitentiaire*, vol. 2, 84–90.

44. Gérard de Nerval, "Mes prisons" (1841). Mentioned in Livio, *Equation That Couldn't Be Solved,* 138.

45. Joseph Liouville, ed., "Oeuvres mathématiques d'Évariste Galois," *Journal de Mathématiques Pures et Appliqués* 9 (October–November 1846): 381–444.

46. Ibid., 382.

47. Ibid., 383.

48. Ibid.

49. Réne Taton, "Galois, Évariste," in Charles C. Gillispie, ed., *Dictionary of Scientific Biography,* vol. 5 (New York: Scribner, 1970), 264.

50. "Évariste Galois," *Magasin Pittoresque* 16 (1848): 227–228.

51. Dupuy, "Vie d'Evariste Galois."

52. A dreyfusard was a supporter of Captain Alfred Dreyfus, the Jewish officer whose conviction for treason sparked the Dreyfus affair that tore apart French society around the turn of the twentieth century. The dreyfusards were generally anticlerical liberals and socialists, and the anti-dreyfusards were conservatives. Dupuy's tomb is inscribed with the words "Un vieux dreyfusard" (an old dreyfusard). The information on Dupuy is derived from the website of the Société Internationale d'Histoire de l'Affaire Dreyfus, http://pockcity.ifrance.com/dupuy.htm, accessed on June 28, 2009.

53. Dupuy, "Vie d'Evariste Galois," 252.

54. Ibid., 251.

55. Évariste Galois, *Oeuvres Mathématiques d'Évariste Galois,* publiées sur les auspices de la Société Mathématique de France avec une introduction par Émile Picard, membre de l'Institut (Paris: Gauthiers-Villars et Fils, 1897).

56. Ibid., vi.

57. Jules Tannery, *Manuscrits d'Évariste Galois* (Paris: Gauthier-Villars 1908), reprinted from *Bulletin des Sciences Mathématiques,* 2nd ser., 30–31 (1906–1907).

58. The speech was published as Jules Tannery, "Discours prononcé à Bourg-la-Reine," *Bulletin des Sciences Mathématiques,* 2nd ser., 33 (1909): 158–164.

59. Ibid., 158.

60. Sarton, "Évariste Galois."

61. Ibid., 256.

62. Ibid., 241.

63. Ibid., 256

64. Ibid.

65. Bell, *Men of Mathematics*, chap. 20.

66. Constance Reid, *The Search for E. T. Bell, Also Known as John Taine* (Washington, DC: Mathematical Association of America, 1993), 289–290.

67. Bell, *Men of Mathematics,* 362.

68. Ibid., 365.

69. Ibid., 362.

70. Tore Frängsmyr, ed., *Les Prix Nobel. The Nobel Prizes 1994* (Stockholm, [Nobel Foundation], 1995); quoted from http://nobelprize.org/nobel_prizes/economics/laureates/1994/nash-autobio.html, accessed June 28, 2009.

71. Rothman, "Genius and the Biographers," discusses the many inaccuracies introduced into Galois' biography when it was transformed into Galois' myth.

72. Bell, *Men of Mathematics,* 369.

73. Petsinis, *French Mathematician,* 421.

4. THE EXQUISITE DANCE OF THE BLUE NYMPHS

1. The information on Abel's life is derived from Oystein Ore, *Niels Henrik Abel, Mathematician Extraordinary* (Minneapolis: University of Minnesota Press, 1957); Oystein Ore, "Abel, Niels Henrik," in *Dictionary of Scientific Biography,* vol. 1, ed. Charles C. Gillispie (New York: Scribner, 1970), 12–17; Arild Stubhaug, *Niels Henrik Abel and His Times: Called Too Soon by Flames Afar,* trans. Richard H. Daly (Berlin: Springer, 2000); and Livio, *Equation That Couldn't Be Solved,* chap. 4.

2. Stubhaug, *Niels Henrik Abel and His Times,* 423.

3. The signatories included the academicians Legendre, Poisson, Sylvestre-François Lacroix, and Jean-Frédéric Théodore Maurice. See Stubhaug, *Niels Henrik Abel and His Times,* 468.

4. The year 1830 was also the one in which Galois submitted his essay to the competition. He was dropped from consideration when Fourier, who was assigned to read his entry, unexpectedly died.

5. Ore, *Niels Henrik Abel,* 200.

6. The text of the academicians' letter to the king can be found in Guillaume Libri, "Abel (Nicholas-Henri)," in *Biographie universelle ancienne et moderne,* vol. 1 (Paris: Michaud, 1843), 59. This biographical dictionary is commonly cited as "Michaud," after its publisher.

7. Stubhaug, *Niels Henrik Abel and His Times,* 27.

8. Ibid., 4.

9. Ibid., 5.

10. Ore, *Niels Henrik Abel,* 237.

11. Ibid.

12. Stubhaug, *Niels Henrik Abel and His Times,* 12.

13. Ore, *Niels Henrik Abel,* 238.

14. Ibid., 238.

15. Ibid., 239–240.

16. Ibid., 242.

17. August Leopold Crelle, obituary of Abel, *Journal für die reine und angewandte Mathematik* (Crelle's journal) 4 (1829): 402–404, reprinted in B. Holmboe, ed., *Oeuvres complètes de N. H. Abel* (Christiania, 1839), xi–xiii.

18. Ibid., xiii.

19. Ore, *Niels Henrik Abel,* 234. "Monumentum aere perennius" (Horace, *Odes* 3.30.1) literally means "a monument more lasting than bronze" and is used to refer to eternal monuments of art.

20. Quoted in Ore, *Niels Henrik Abel,* 235.

21. Frédéric Saigey, obituary of Abel, *Annales des Sciences d'Observations,* May 1829; quoted in Ore, *Niels Henrik Abel,* 153–154, 252–253.

22. Libri presented his biography to the academy in 1833, and it was published shortly thereafter in Michaud's biographical dictionary, *Biographie universelle ancienne et moderne* (Paris: Michaud, 1834), vol. 56, supplement, 22–29, and reprinted in later editions. The following quotes are from the 1843 edition, 1:58–62. See also Ore, *Niels Henrik Abel,* 249–251; and Stubhaug, *Niels Henrik Abel and His Times,* 550–551.

23. Michaud, *Biographie universelle ancienne et moderne* (1843), 1:59. Abel did spend a few weeks in the summer of 1827 with his mother in Lunde, although it is not clear whether Libri knew that. One wonders whether Libri is here confusing Abel with Galois, who according to Sophie Germain's letter to Libri had to move in with his mother.

24. Michaud, *Biographie universelle ancienne et moderne* (1843), 1:59.

25. Ibid.

26. Ibid., 1:60.

27. Ore, *Niels Henrik Abel,* 261; Stubhaug, *Niels Henrik Abel and His Times,* 551.

28. Norway at the time of Abel was under Swedish suzerainty, and it is therefore reasonable of Raspail to refer to Abel as a Swede.

29. The speech by Raspail is quoted in Ore, *Niels Henrik Abel,* 255–257.

30. The material on the Abel centennial is from Nils Voje Johansen and Yngvar Reichelt, eds., "The Abel Centennial 1902 and Bicentennial 2002," trans. Olav Arnfinn Laudal and Ragni Pienne, in Olav Arnfinn Laudal and Ragno Piene, eds.,

The Legacy of Niels Henrik Abel: The Abel Bicentennial, Oslo 2002 (CD-ROM, Berlin: Springer-Verlag, 2004).

31. Niels Henrik Abel, *Oeuvres complètes de Niels Henrik Abel,* ed. Sophus Lie and Ludwig Sylow (Christiania: Grøndahl & Søn, 1881).

32. "Speech of Prof. Dr. P. L. M. Sylow," in Johansen and Reichelt, "Abel Centennial 1902 and Bicentennial 2002."

33. Ibid.

34. Speech of R. A. Forsyth," in Johansen and Reichelt, "Abel Centennial 1902 and Bicentennial 2002."

35. Ibid.

36. Speech of Emile Picard, in Johansen and Reichelt, "Abel Centennial 1902 and Bicentennial 2002."

37. "Niels Henrik Abel," a cantata by Bjørnstjerne Bjørnson, music by Christian Sinding, in Johansen and Reichelt, "Abel Centennial 1902 and Bicentennial 2002."

38. Stubhaug, *Niels Henrik Abel and His Times.* The book was first published in Norwegian as *Et foranskutt lynn: Niels Henrik Abel og hans tid* (Oslo: H. Aschenhoug & Co., 1996).

39. In 2002, on the bicentenary of Abel's birth, the Norwegian government acted on a plan that had first been proposed in the 1902 centennial celebrations, establishing the Abel Prize for outstanding scientific work in the field of mathematics. It is one of the most prestigious international prizes in the field.

40. Galois, "Préface," in Bourgne and Azra, *Écrits et mémoires mathématiques d'Évariste Galois,* 5.

41. Bell, *Men of Mathematics,* 362.

42. Ore, *Niels Henrik Abel,* 253–254; Stubhaug, *Niels Henrik Abel and His Times,* 415–417; Livio, *Equation That Couldn't Be Solved,* chaps. 4, 5.

5. A MARTYR TO CONTEMPT

1. Jean-Victor Poncelet, *Applications d'analyse et de géométrie,* vol. 2 (Paris, 1864), 564; quoted in Bruno Belhoste, *Augustin-Louis Cauchy: A Biography,* trans. Frank Ragland (New York: Springer-Verlag, 1991), 55.

2. The term *ultra* is short for *ultramontain,* referring to this party's insistence on the pope's supremacy in the French Catholic Church.

3. The biographical information on Cauchy is based on Belhoste, *Augustin-Louis Cauchy.*

4. Ibid., 1–8, 224.

5. Ibid., 22.

6. Ibid., 40.

7. On the centrality of the École polytechnique for French mathematical sciences in the nineteenth century, see Bruno Belhoste, "The École Polytechnique and Mathematics in Nineteenth-Century France," in Umberto Bottazzini and Amy Dahan Dalmedico, eds., *Changing Images in Mathematics from the French Revolution to the New Millennium* (London: Routledge, 2001), 15–30.

8. Belhoste, *Augustin-Louis Cauchy,* 46–47.

9. For more on Gaspard Monge and his legacy at the École polytechnique, see Belhoste, "École Polytechnique and Mathematics," 22.

10. On the prevailing attitudes toward mathematics and its applications at the École polytechnique, see Belhoste, "École Polytechnique and Mathematics." The focus on the practical uses of mathematics for engineering and industrial development became official state policy in France after the Revolution, and the École polytechnique was founded to put this policy into effect. See Margaret C. Jacob and Larry Stewart, *Practical Matter: Newton's Science in the Service of Industry and Empire* (Cambridge, MA: Harvard University Press, 2004), 138–145; and Jeff Horn and Margaret C. Jacob, "Jean-Antoine Chaptal and the Cultural Roots of French Industrialization," *Technology and Culture* 39, no. 4 (1998): 671–698.

11. On the difference between eighteenth- and nineteenth-century mathematics, and on Cauchy's central role in this transition, see Judith V. Grabiner, *The Origins of Cauchy's Rigorous Calculus* (New York: Dover, 2005; first published, Cambridge, MA: MIT Press, 1981), esp. chaps. 1 and 2. The transformation of the standards and practice of mathematics in the early nineteenth century was so profound that it has been called "a second rebirth of the subject—its first birth having occurred among the ancient Greeks." See Howard Stein, "*Logos,* Logic, and *Logistiké:* Some Philosophical Remarks on Nineteenth-Century Transformation of Mathematics," in William Aspray and Philip Kitcher, eds., *History and Philosophy of Modern Mathematics* (Minneapolis: University of Minnesota Press, 1988), 238.

12. On the practical emphasis in French scientific education after the Revolution, see Margaret C. Jacob, *Scientific Culture and the Making of the Industrial West* (Oxford: Oxford University Press, 1997), esp. chap. 8, "French Industry and Engineers under Absolutism and Revolution."

13. Belhoste, *Augustin-Louis Cauchy,* 65.

14. A. L. Cauchy, *Cours d'analyse de l'École royale polytechnique* (Paris, 1821).

15. For more on the contents of *Cours d'analyse,* see Ivan Grattan-Guinness, *Convolutions in French Mathematics, 1800–1840* (Berlin: VEB, 1990), 2:712–727.

16. On Cauchy's troubles at the École polytechnique with both students and faculty, see ibid., 2:709–712.

17. Belhoste, *Augustin-Louis Cauchy,* 72.

18. Ibid., 73.

19. A. L. Cauchy, *Résume des leçons données à l'École royale polytechnique sur le calcul infinitésimal* (Paris: Imprimerie Royale, 1823).

20. Belhoste, *Augustin-Louis Cauchy,* 81.

21. The delta-epsilon method is a rigorous mathematical expression of the intuition that even if a function is not defined at a point, it can nevertheless approach its limit value at that point as closely as one wishes. That is, if the function f is undefined at the point x_0 but does have a limit there, then for every ε, no matter how small, there is a value δ such that if $|x - y| < \delta$, then $|f(x) - f(x_0)| < \varepsilon$.

22. Belhoste, *Augustin-Louis Cauchy,* 82.

23. On Prony's criticism of Cauchy's teaching, see Grattan-Guinness, *Convolutions in French Mathematics,* 2:800–803.

24. Belhoste, *Augustin-Louis Cauchy,* 84.

25. Ibid.

26. Ibid., 84–85.
27. Ibid., 138.
28. Ibid.
29. Ibid., 140.
30. On Cauchy's exile, see ibid., chaps. 9–10.
31. Ibid., 148.
32. On Cauchy's sojourn as tutor to the Duke of Bordeaux, see ibid., chap. 10.
33. On Cauchy's candidacy for a position in the Bureau des longitudes, see ibid., 174–177.
34. Ibid., 184–186.
35. Ibid., 223–225. Under the Second Empire, which was established in 1852, Napoleon III did require an oath of loyalty from civil servants. The emperor, however, explicitly exempted Cauchy from this requirement, and he held on to his position at the Faculté des Sciences. See ibid., 228.
36. Ibid., 224–227.
37. A. L. Cauchy, "Recherche de la vérité," lecture given on April 14, 1842, printed in *Bulletin de l'Institut Catholique*, 2nd installment, 1842, 19–28. The title was likely inspired by the work of the Catholic Cartesian philosopher Nicolas Malebranche (1638–1715), whose most important work is titled *De la recherche de la vérité.*
38. Cauchy, "Recherche de la Verite," 20.
39. Ibid.
40. Ibid., 21.
41. Ibid., 23.
42. Ibid., 28.
43. A. L. Cauchy, "Sur quelques préjugés contre les physiciens et les géomètres," lecture given on March 4, 1842, printed in *Bulletin de l'Institut Catholique,* 1st installment, 1842, 43.
44. Ibid.
45. Ibid., 45.

6. THE POETRY OF MATHEMATICS

1. Galois' poem and the tale of its discovery by Chevalier conclude Dupuy's biography of Galois, "Vie d'Evariste Galois."
2. The account of Romanticism here is based on Hugh Honour, *Romanticism* (New York: Harper and Row, 1979), as well as Hans Georg Artur Viktor Schenk, *The Mind of the European Romantics: An Essay in Cultural History* (London: Constable, 1966).
3. Literary theorist George Steiner suggests that Romanticism was not a reaction to the horrors of the revolutionary era but rather to the social and political repression that followed in the decades after 1815. Be that as it may, there is no question that Romanticism defined itself as the counterpoint to the eighteenth-century "age of reason." See George Steiner, *In Bluebeard's Castle: Some Notes towards the Redefinition of Culture* (New Haven: Yale University Press, 1971), chap. 1, "The Great Ennui."

4. On the romantic ideal of the tragic life of artists, poets, and musicians, see Honour, *Romanticism,* esp. chap. 7, "Artist's Life."

5. Heinrich Heine, quoted in Honour, *Romanticism,* 256–258.

6. Sarton, "Évariste Galois," 241.

7. Ibid., 256.

8. Nathaniel Hawthorne, quoted in Honour, *Romanticism,* 254.

9. The term *mathematical persona* is inspired by Lorraine Daston and Peter Galison's use of the term *scientific persona* in *Objectivity* (New York: Zone Books, 2007), 216–233.

10. On the shifting of the center of mathematical research from France to Germany in the nineteenth century, see Joseph Dauben, "Mathematics in Germany and France in the Early 19th Century: Transmission and Transformation," in Hans Niels Jahnke and M. Otte, eds., *Epistemological and Social Problems of the Sciences in the Early 19th Century* (Dordrecht: D. Reidel, 1981), 371–399; Ivor Grattan-Guinness, "The End of Dominance: The Diffusion of French Mathematics Elsewhere, 1829–1870," in Karen Hunger Parshall and Adrian C. Rice, eds., *Mathematics Unbound: The Evolution of an International Mathematical Research Community, 1800–1945* (Providence, RI: American Mathematical Society, 2002), 17–44; Gert Schubring, "The German Mathematical Community," in J. Fauvel, R. Flood, and R. Wilson, eds., *Möbius and His Band* (Oxford: Oxford University Press, 1993), 21–33; and more generally, H. G. W. Begehr, H. Koch. J. Kramer, N. Schappacher, and E. J. Thiele, eds., *Mathematics in Berlin* (Berlin: Birkhäuser Verlag, 1998), 9–48.

11. For a recent fictionalized account of Ramanujan as a tragic romantic hero, see David Leavitt, *The Indian Clerk* (New York: Bloomsbury USA, 2007).

12. This account of Gödel's life and work in based on Rebecca Goldstein, *Incompleteness: The Proof and Paradox of Kurt Gödel* (New York: W. W. Norton, 2005).

13. Sylvia Nasar, *A Beautiful Mind* (New York: Touchstone Books, 1994). The movie of the same name, directed by Ron Howard and starring Russell Crowe as Nash, was released in 2001. Nasar's book is the most successful mathematical biography in recent years. She has become a modern-day E. T. Bell, communicating the stories, as well as the legends, of mathematics to a broad audience.

14. Nasar, *A Beautiful Mind,* 12.

15. Sylvia Nasar and David Gruber, "Manifold Destiny," *New Yorker,* August 28, 2006.

16. Phillip Griffiths, quoted in Nasar and Gruber, "Manifold Destiny."

17. G. H. Hardy, *A Mathematician's Apology* (Cambridge: Cambridge University Press, 2004; first published in 1940).

18. Ibid., 84.

19. Ibid., 85.

20. Ibid., 139.

21. C. P. Snow, foreword to Hardy, *A Mathematician's Apology,* 50–51.

22. Terrall, *Man Who Flattened the Earth,* 355.

23. Joseph Fourier, introduction to *Théorie analytique de la chaleur* (Paris, 1822); quoted in Winfried Scharlau, "The Origins of Pure Mathematics," in Jahnke

and Otte, *Epistemological and Social Problems of the Sciences in the Early Nineteenth Century,* 346.

24. Fourier, introduction to *Théorie analytique de la chaleur;* quoted in Scharlau, "Origins of Pure Mathematics," 344.

25. See the account in Herbert Pieper, "Carl Gustav Jacobi," in Begehr et al., eds., *Mathematics in Berlin,* 46.

26. Pieper, "Carl Gustav Jacobi," 46, quotes this passage but translates it as the "honor of the human mind" rather than the "honor of the human spirit." The French original reads "le but unique de la science, c'est l'honneur de l'esprit humain," and there seems no reason to translate "esprit" as "mind." See the quote in Scharlau, "Origins of Pure Mathematics," 337, and Helmut Pulte, "Jacobi's Criticism of Lagrange: The Changing Role of Mathematics in the Foundations of Classical Mechanics," *Historia Mathematica* 25 (1998): 162.

27. Carl Gustav Jacobi, quoted in John Fauvel, Raymond Flood, and Robin Wilson, eds., *Möbius and His Band* (Oxford: Oxford University Press, 1993), 29.

28. Carl Gustav Jacobi, quoted in Pulte, "Jacobi's Criticism of Lagrange," 178.

29. Auguste Crelle, quoted in Fauvel, Flood, and Wilson, *Möbius and His Band,* 31.

30. Abel to Holmboe, 1826; quoted in Jesper Lützen, "The Foundation of Analysis in the 19th Century," in Hans Niels Jahnke, ed., *A History of Analysis* (Providence, RI: American Mathematical Society, 2003; first published in German in 1999), 177.

31. Abel to Hansteen, March 29, 1826; quoted in Lützen, "Foundation of Analysis in the 19th Century," 176.

32. Niels Henrik Abel, quoted in Lützen, "Foundation of Analysis in the 19th Century," 177.

33. The quote is from Cauchy, "Recherche de la Verite," 21. For a discussion of the radical break in mathematics between the eighteenth and nineteenth centuries, see Jeremy Gray, "The Nineteenth-Century Revolution in Mathematical Ontology," in Donald Gillies, ed., *Revolutions in Mathematics* (Oxford: Clarendon Press, 1992), 226–248. Gray calls the eighteenth-century approach "naïve abstractionism." See also Jesper Lützen, "Between Rigor and Application," in *The Cambridge History of Science,* vol. 5, ed. Mary Jo Nye (Cambridge: Cambridge University Press, 2002), 468–487. Philosopher Howard Stein called nineteenth-century mathematics "a second rebirth of the subject" in "*Logos,* Logic, and *Logistiké,*" 238.

34. Hardy, *A Mathematician's Apology,* 135.

7. PURITY AND RIGOR

1. Cauchy, introduction to *Cours d'analyse*, iij.

2. Ibid., vij.

3. This is not to say that Cauchy was uninterested in new results. To the contrary, over the course of his long career Cauchy made innumerable innovations in both mathematics and what is known today as mathematical physics. But his most enduring contribution to academic mathematics was his restructuring of analysis

as a self-contained and rigorous edifice. Cauchy's *Cours d'analyse* of 1821 and *Résumé des leçons* of 1823, in which he outlined his approach, were highly unpopular among the teachers and students of the Ecole polytechnique for whom they were intended (see Chapter 5), but they provide the framework for the teaching of analysis in academic mathematics departments to this day.

4. On Cauchy's definition of function, see Lützen, "Foundation of Analysis in the 19th Century," 156–158.

5. Euler in fact defined the term *function* differently in different places, sometimes in ways quite similar to Cauchy's definition. In his mathematical practice, however, he stuck to the definition of a function as an analytic expression, whereas Cauchy founded his system on his very different definition of a function. See Lützen, "Foundation of Analysis in the 19th Century," 156.

6. On eighteenth-century mathematicians' faith in the intrinsically mathematical nature of the world, see Chapter 2, as well as Pulte, "Jacobi's Criticism of Lagrange," 154–184, esp. 156.

7. On the definition of the term *variable,* see Cauchy, *Cours d'analyse,* 19; quoted in Lützen, "Foundation of Analysis in the 19th Century," 158. On Cauchy's definition of a function in the *Cours d'analyse,* see Lützen, "Foundation of Analysis in the 19th Century," 156.

8. Cauchy, *Cours d'analyse,* 19; quoted in Lützen, "Foundation of Analysis in the 19th Century," 158, and with a slightly different translation in Judith C. Grabiner, *The Origins of Cauchy's Rigorous Calculus* (New York: Dover, 2005), 7.

9. Jean d'Alembert, "Limite," in Denis Diderot and Jean d'Alembert, eds., *Encyclopédie,* vol. 9 (Paris: Le Breton, 1765), 542.

10. "They are neither finite quantities nor quantities infinitely small, nor yet nothing. May we not call them the ghosts of departed quantities?" George Berkeley, *The Analyst: A Discourse Addressed to an Infidel Mathematician* (London, 1634), aphorism 35. For a thorough discussion of eighteenth-century critiques of the calculus and responses to them, see Carl B. Boyer, *The History of the Calculus and Its Conceptual Development* (New York: Dover Publications, 1959), chap. 6.

11. On the difference between Cauchy's concept of the limit and that of his eighteenth-century predecessors, see Grabiner, *Origins of Cauchy's Rigorous Calculus,* chap. 4, esp. 84–86.

12. Cauchy, *Cours d'analyse,* 19; quoted in Lützen, "Foundation of Analysis in the 19th Century," 158.

13. Cauchy, *Résumé des leçons,* 22; quoted in Lützen, "Foundation of Analysis in the 19th Century," 159.

14. On Cauchy's sources, see Grabiner, *Origins of Cauchy's Rigorous Calculus,* chaps. 4, 5.

15. On the continuation of the Enlightenment tradition in French mathematics, see Lorraine Daston, "The Physicalist Tradition in Early 19th Century French Geometry," *Studies in the History and Philosophy of Science* 17, no. 3 (1986): 269–295. On the leading role in France of graduates of the École polytechnique, see Bruno Belhoste, *La formation d'une technocratie: L'École polytechnique et ses élèves de la révolution au Second Empire* (Paris: Belin, 2003).

16. The Cossists themselves would have presented both the problem and the solution quite differently, but my focus here is on how the problem was viewed by eighteenth- and nineteenth-century mathematicians. They looked back at the work of their predecessors from antiquity to the Renaissance and translated it into modern algebraic notation. For more on the Cossists, see Michael S. Mahoney, *The Mathematical Career of Pierre de Fermat* (Princeton, NJ: Princeton University Press, 1973), 4–7; Peter Pesic, *Abel's Proof: An Essay on the Sources and Meaning of Mathematical Unsolvability* (Cambridge, MA: MIT Press, 2003), 30–41; and John Derbyshire, *Unknown Quantity: A Real and Imaginary History of Algebra* (Washington, DC: Plume, 2006), 72–80.

17. The following account of the discovery of the solution to the cubic by del Ferro, Tartaglia, and Cardano is derived primarily from Derbyshire, *Unknown Quantity,* 74–80. See also Pesic, *Abel's Proof,* 32–40.

18. Girolamo Cardano, *Artis magnae sive de regulis algebraicis liber unus* (Nuremberg, 1545).

19. For the Cossist terminology and notation, see Mahoney, *Mathematical Career of Pierre de Fermat*, 34–36.

20. This discussion of the general solution to the cubic follows Derbyshire, *Unknown Quantity,* 57–64. For proofs of the formulas, see Derbyshire, *Unknown Quantity,* 57–64, as well as Pesic, *Abel's Proof,* 23–46.

21. Joseph Louis Lagrange, "Réflexions sur la resolution algébrique des équations," in *Oeuvres de Lagrange,* vol. 3 (Paris: Gauthier-Villars, 1869), 205–516. This work was first published in the *Nouveaux Mémoires de l'Academie Royale des Sciences et Belles-Lettres* in 1771.

22. The following presentation of Lagrange's and Vandermonde's theory of equations is based on Harold M. Edwards, *Galois Theory* (New York: Springer, 1984), 17–26; Derbyshire, *Unknown Quantity*, 117–125; Pesic, *Abel's Proof,* chap. 5; Reinhard Laubenbacher and David Pengelley, *Mathematical Expeditions: Chronicles by Explorers* (New York: Springer, 1999), 233–347; and Lagrange, "Réflexions" 205–421.

23. Lagrange, "Réflexions," 206.

24. Ibid.

25. Ibid.

26. Lagrange in fact used α, not ω, to denote the cube root of unity, but I use ω here to preserve consistency with the previous discussion.

27. As mentioned earlier, $\omega^2 + \omega + 1 = 0$ because ω is a cube root of 1 and therefore a solution to the equation $x^3 - 1 = 0$. Now $x^3 - 1 = (x - 1)(x^2 + x + 1)$, where ω and ω^2 are the roots of $x^2 + x + 1$, which means that $\omega^2 + \omega + 1 = 0$. In the expansion of the resolvent equation $f(X)$ it is found that all the elements except X^3, t_1^3, and t_4^3 are multiplied by $\omega^2 + \omega + 1$ and are therefore zero.

28. A symmetric polynomial is one that remains unchanged under any permutation of its variables. The polynomial $xy + yz + xz$, for example, is symmetric, while the polynomial $xy + 2yz + 3zx$ is not. For more on symmetric polynomials and the ways symmetric polynomials in the roots of an equation can be expressed in terms of the coefficients, see Edwards, *Galois Theory,* 6–13, and Derbyshire, *Unknown Variable*, 97–104.

29. The first challenge in determining x, y, and z from the six t_ks is to identify t_1 and t_4. Vandermonde settled for trial and error in this task, whereas Lagrange offered a method. Once these are known, then

$$x = \frac{1}{3}[(x + y + z) + t_1 + t_4];$$

$$y = \frac{1}{3}[(x + y + z) + \omega^2 t_1 + \omega t_4];$$

and

$$z = \frac{1}{3}[(x + y + z) + \omega t_1 + \omega^2 t_4].$$

Because $x + y + z = -p$, these values are easily derived. On why this is so, see Edwards, *Galois Theory,* 17–20.

30. On solving the quadratic by this method, see Edwards, *Galois Theory,* 17–18; and Derbyshire, *Unknown Quantity,* 117–118.

31. A bicubic equation is an equation of the fourth degree that is in fact a quadratic equation of x^2. It takes the form $x^4 + px^2 + q = 0$, which can be solved for x^2 and thereby for x.

32. See Pesic, *Abel's Proof,* chap. 6; Stubhaug, *Niels Henrik Abel and His Times,* 238–239, 302–303; Ore, *Niels Henrik Abel,* 34–45, 69–70; and Derbyshire, *Unknown Quantity,* 128–132. Abel was not in fact the first to prove that the quintic was not solvable by radicals. Unknown to Abel, the Italian Paolo Ruffini had published six versions of his proof of this result between 1799 and 1813. But Ruffini's proof was written in an obscure style that made it difficult to evaluate, and it was ultimately judged to lack a crucial component. As a result, credit for the proof usually goes to Abel, although the theorem is sometimes known as the Abel-Ruffini theorem. On Ruffini, see Pesic, *Abel's Proof,* 80–83; and Derbyshire, *Unknown Quantity,* 125–127.

33. More precisely, he was searching for the characteristics that make any equation of the form $P(x) = 0$ solvable, where $P(x)$ is a polynomial.

34. The technical exposition that follows relies on Edwards, *Galois Theory,* and chapter 5 of Laubenbacher and Pengelley, *Mathematical Expeditions,* as well as on Galois' memoir to the Paris Academy, included as appendix 1 in Edwards, *Galois Theory.*

35. The term *field* was coined later in the nineteenth century by Richard Dedekind and is not used by Galois, who simply talks about sets of values that can all be expressed as "rational functions" of each other. This is very close to the modern definition of a field that I use here. A rational function is one that makes use only of addition, subtraction, multiplication, and division, and whose general form is therefore the division of one polynomial by another.

36. Not everyone was satisfied with the proofs that Galois offered for his results, although no one ultimately disputed their correctness. When Poisson reviewed Galois' memoir for the Paris Academy in 1831, he noted in the margins that one of Galois' "proofs" was far too sketchy and preliminary, even though the theorem itself was "true by article 100 of the memoir of Lagrange, Berlin, 1771." When he finally received the manuscript with the comments back from the academy, the resentful Galois wrote his response next to Poisson's comments: "On

jugera," meaning "one will judge for himself," or perhaps "we shall see"; quoted in Edwards, *Galois Theory,* 43.

37. The requirement that the equation have n different roots is not a limitation on the generality of Galois' work, because it can be shown that an equation whose solutions include double roots can be replaced by one with n distinct roots.

38. The term $n!$, called "n factorial," is defined as $1 \times 2 \times 3 \times \ldots \times (n-1) \times n$.

39. Galois carefully defined the terms of his criterion, although I will not do so here. Roughly speaking, a *group* is a set of substitutions that transform a given arrangement of objects into a set of other arrangements in such a way that no matter which arrangements one starts with, the substitutions in the group will produce all other arrangements. A *subgroup* is a subset of the full group that is itself a group. Each subgroup can be presented in different ways, depending on the arrangement it takes as its starting point. A *normal subgroup* is one in which the different presentations can be derived from one another by a single substitution. Note that although Galois was the first to use the term *group* in this way, the term *normal subgroup* is of later provenance. Nevertheless, Galois did recognize the importance of this particular type of subgroup. See Edwards, *Galois Theory,* 50.

40. Laubenbacher and Pengelley, *Mathematical Expeditions,* 247.

41. Galois, "Discours préliminaire," September 1830, in Bourgne and Azra, *Écrits et mémoires mathématiques d'Évariste Galois,* 39–41; also quoted in Laubenbacher and Pengelley, *Mathematical Expeditions,* 254–255.

42. Livio, *Equation That Couldn't Be Solved,* 133.

43. Galois, "Discours préliminaire." 41.

44. On Jacobi, see Fauvel, Flood, and Wilson, *Möbius and His Band,* 29.

45. Galois, "Préface," December 1831, in Bourgne and Azra, *Écrits et mémoires mathématiques d'Évariste Galois,* 9–11; John Fauvel and Jeremy Gray, *The History of Mathematics: A Reader* (London: Macmillan, 1987), 505; Laubenbacher and Pengelley, *Mathematical Expeditions,* 217.

46. Galois, "Préface," 9; Fauvel and Gray, *History of Mathematics,* 505; Laubenbacher and Pengelley, *Mathematical Expeditions,* 217.

47. Galois, "Préface," 9–11; Fauvel and Gray, *History of Mathematics,* 505; Laubenbacher and Pengelley, *Mathematical Expeditions,* 217. The quote is from page 11 of Galois' "Préface."

48. J. Dieudonné, preface to Bourgne and Azra, *Écrits et mémoires mathématiques d'Évariste Galois,* v, vi.

49. This short list of the pioneering mathematicians of the early nineteenth century is far from definitive and includes only those who have already been discussed. Others who can be included are Bernard Bolzano and Gauss in his unpublished works, as well as János Bolyai and Nikolai Lobachevskii, who will be discussed in Chapter 8.

8. THE GIFTED SWORDSMAN

1. From Christopher Clavius, "In disciplinas mathematicas prolegomena," in Clavius, *Opera mathematica,* vol. 1 (Mainz, 1611), quoted in James A. Lattis,

Between Copernicus and Galileo: Cristoph Clavius and the Collapse of Ptolemaic Astronomy (Chicago: University of Chicago Press, 1994), 35.

2. Euclid, *Euclid's Elements: All Thirteen Books Complete in One Volume,* ed. Dana Densmore, trans. Thomas L. Heath (Santa Fe: Green Lion Press, 2002), 2.

3. Ibid.

4. On the efforts to prove the fifth postulate and on the history of non-Euclidean geometry in general, see Roberto Bonola, *Non-Euclidean Geometry,* trans. H. S. Carslaw (New York: Dover Publications, 1955; first published in Italian in 1912); Jeremy J. Gray, *Ideas of Space: Euclidean, Non-Euclidean, and Relativistic* (Oxford: Oxford University Press, 1979); Jeremy J. Gray, *János Bolyai, Non-Euclidean Geometry, and the Nature of Space* (Cambridge, MA: Burndy Library Publications, 2006); B. A. Rosenfeld, *A History of Non-Euclidean Geometry: Evolution of the Concept of a Geometric Space,* trans. Abe Shenitzer (New York: Springer-Verlag, 1988; first published in Russian in 1976); and Harold E. Wolfe, *Non-Euclidean Geometry* (New York: Dryden Press, 1945).

5. For more on the ancient critics of the fifth postulate, see Bonola, *Non-Euclidean Geometry,* 1–9.

6. Gray, *Ideas of Space,* 52.

7. Amir R. Alexander, *Geometrical Landscapes* (Stanford: Stanford University Press, 2002), chap. 6.

8. George Bruce Halsted, ed. and trans., *Girolamo Saccheri's Euclides Vindicatus* (Chicago: Open Court, 1920). First published as Hieronymo Saccherio, *Euclides ab omni naevo vindicatus, sive conatus geometricus quo stabiliuntur prima ipsa universae geometriae principia* (Milan: Paolo Antonio Montano, 1733).

9. Saccheri was not the first to draw attention to this quadrilateral. It had been used some years earlier by the Italian mathematician Giordano Vitale (1633–1711). See Gray, *Ideas of Space,* 52–54.

10. Quoted in Gray, *Ideas of Space,* 58.

11. Halsted, *Girolamo Saccheri's Euclides Vindicatus,* 15; Gray, *Ideas of Space,* 61–62.

12. Alexis Clairaut, quoted in Gray, *János Bolyai,* 27.

13. Gray, *János Bolyai,* 28.

14. Bonola, *Non-Euclidean Geometry,* 53–54.

15. On Lambert's astronomical views and his theory of the Milky Way, see Michael J. Crowe, *The Extraterrestrial Life Debate, 1750–1900* (Mineola, NY: Dover Publications, 1999; first published, Cambridge: Cambridge University Press, 1986), 55–59.

16. On Lambert's work on the parallels, see Bonola, *Non-Euclidean Geometry,* 44–51; Gray, *Ideas of Space,* 63–68; and Gray, *János Bolyai,* 41–42.

17. Saccheri based his definitions on a quadrilateral with two right angles and two equal sides; Lambert preferred a quadrilateral with three right angles and the fourth one either right, acute, or obtuse.

18. Bonola, *Non-Euclidean Geometry,* 50.

19. On French mathematicians' disinterest in the problem of the parallel postulate and non-Euclidean geometry, see Gray, *Ideas of Space,* 69, 74–76.

20. On Lambert's enthusiasm for the plurality of worlds, see Crowe, *Extraterrestrial Life Debate,* 55–59.

21. On Gauss's belief in the plurality of worlds, see Crowe, *Extraterrestrial Life Debate,* 204–208.

22. Gauss, quoted in Gray, *János Bolyai,* 44–45.

23. Gauss to Friedrich Wilhelm Bessel, January 27, 1829; quoted in Bonola, *Non-Euclidean Geometry,* 67; Gray, *Ideas of Space,* 76; and Gray, *János Bolyai,* 45. On Gauss's reluctance to make his views public, see Bonola, *Non-Euclidean Geometry,* 66–67.

24. Gray, *János Bolyai,* 43–48.

25. Bonola, *Non-Euclidean Geometry,* 62–63; Gray, *János Bolyai,* 46.

26. Bonola, *Non-Euclidean Geometry,* 77; Gray, *Ideas of Space,* 87.

27. Bonola, *Non-Euclidean Geometry,* 76–77.

28. Ibid., 76.

29. Ibid., 77–83.

30. Wolfgang (Farkas) Bolyai, *Tentamen juventutem studiosam in elementa matheseos puræ elementaris ac sublimioris methods intuitiva evidentiaque huic propria introducendi, cum appendice triplici* (Marosvásárhely, 1832).

31. H. Meschkowski, *Noneuclidean Geometry,* trans. A. Shenitzer (New York: Academic Press, 1964), 31; Fauvel and Gray, *History of Mathematics,* 527.

32. Farkas Bolyai, quoted in D. J. Struik, "János Bolyai," in *Dictionary of Scientific Biography,* vol. 2 (New York: Scribner, 1970), 270.

33. Farkas Bolyai to János Bolyai, quoted in Gray, *János Bolyai,* 51–52. Slightly different translations of these quotes are included in Fauvel and Gray, *History of Mathematics,* 528; Gray, *Ideas of Space,* 96; and Meschkowski, *Noneuclidean Geometry,* 33.

34. Franz Schmidt, "Notice sur la vie et les travaux de W. et J. Bolyai," in Jules Hoüel, ed., *La science absolue de l'espace, indépendente de la vérité ou la fausseté de l'axiome XI d'Euclide (que l'on ne pourra jamais établir a priori); par Jean Bolyai, capitaine au Corps de Genie dans l'armée autrichienne* (Paris: Gauthier-Villars, 1868), 20.

35. János Bolyai to Farkas Bolyai, November 3, 1823; quoted in Gray, *János Bolyai,* 52; also quoted with variations in Bonola, *Non-Euclidean Geometry,* 98; and Meschkowski, *Noneuclidean Geometry,* 33.

36. For more on Gauss's views on geometry, see Luciano Boi, "The 'Revolution' in the Geometrical Vision of Space in the Nineteenth Century, and the Hermeneutical Epistemology of Mathematics," in Gillies, *Revolutions in Mathematics,* 183–208, esp. 203–206.

37. Schmidt, "Notice sur la vie et les travaux de W. et J. Bolyai," 8–18.

38. János Bolyai to Farkas Bolyai, November 3, 1823; quoted in Gray, *János Bolyai,* 52.

39. Carl Friedrich Gauss to Farkas Bolyai, March 6, 1832; quoted in Gray, *János Bolyai,* 53–54. A different translation is given in Bonola, *History of Non-Euclidean Geometry,* 100. For a more complete translation, see János Bolyai, *Appendix,* ed. Ferenc Kárteszi (Amsterdam: North-Holland, 1987) 34–35. The original is in Carl Friedrich Gauss, *Werke,* vol. 8 (Leipzig: B. G. Teubner, 1900), 220–224.

40. Gauss, quoted in Tord Hall, *Carl Friedrich Gauss,* trans. Albert Froderberg (Cambridge, MA: MIT Press, 1970), 138.

41. Quoted in Ore, *Niels Henrik Abel,* 155–156.

42. Schmidt, "Notice sur la vie et les travaux de W. et J. Bolyai," 20.

43. For more on János Bolyai's mathematical work in his later years, see Elemér Kiss, *Mathematical Gems from the Bolyai Chests,* trans. Anikó Csirmaz and Gábor Oláh (Budapest: Akádemiai Kiadó and TypoTEX, 1999).

44. The notion that Bolyai's and Lobachevskii's non-Euclidean geometry marks a revolutionary step in the evolution of mathematics is common in the literature on the history of mathematics. See, for example, Yuxin Zheng, "Non-Euclidean Geometry and Revolutions in Mathematics," in Gillies, *Revolutions in Mathematics,* 169–182; and Boi, "'Revolution' in the Geometrical Vision of Space in the Nineteenth Century," 183–208. Zheng, "Non-Euclidean Geometry," 176, includes the following quote from mathematician and historian of mathematics Morris Kline: "Amidst all the complex technical creations of the nineteenth century the most profound one, non-Euclidean geometry, was technically the simplest. This creation gave rise to important new branches of mathematics, but its most significant implication was that it obliged mathematicians to revise radically their understanding of the nature of mathematics and its relation to the physical world."

45. The following exposition of Bolyai's geometry is based on the excellent account in Gray, *János Bolyai,* 55–68, with additional information from Gray, *Ideas of Space,* 96–112, and Bonola, *Non-Euclidean Geometry,* 96–113.

46. After figure 4 in Gray, *János Bolyai,* 55.

47. After figure 5, ibid., 56.

48. After figure 6, ibid., 57.

49. After figure 7, ibid., 58.

50. See Gray, *János Bolyai,* 66. For slightly different formulations of the same relationship, see Gray, *Ideas of Space,* 107; and Bonola, *Non-Euclidean Geometry,* 90. "Sinh" denotes the hyperbolic sine function, and "cot" denotes the cotangent.

51. János Bolyai to Farkas Bolyai, November 3, 1823; quoted in Gray, *János Bolyai,* 52.

52. Eric Temple Bell's brief account of Lobachevskii's life in *Men of Mathematics* makes it clear that he, at least, saw the Russian in precisely these tragic terms. See E. T. Bell, "The Copernicus of Geometry: Lobatchewsky," in *Men of Mathematics,* 294–306.

53. See letter from Abel to Hansteen, March 29, 1826; quoted in Lützen, "Foundation of Analysis in the 19th Century," 176.

54. D'Alembert, *Preliminary Discourse,* 18–20.

CONCLUSION

1. On Abel and Gørbitz, see Stubhaug, *Niels Henrik Abel and His Times,* 395–396.

2. On Gros as an early romantic painter, see Honour, *Romanticism,* 35–39; for Gros' influence on romantic painters, see ibid., 39–55.

3. Another famous portrait in this tradition is Eugène Delacroix's (1798–1863) portrait of his fellow painter the Baron Schwiter, a work begun in the same year in which Abel was in Paris and exhibited in the Salon of 1827.

4. "Évariste Galois," *Magasin Pittoresque* 16 (1848): 227–228. The portrait is on page 228.

5. Dupuy, "Vie d'Evariste Galois." The portrait is inserted after page 200.

6. János Bolyai, *Appendix*, 232.

7. In their 2007 book *Objectivity*, Lorraine Daston and Peter Galison argue that in the nineteenth century science and the creative arts parted ways. Science was henceforth concerned with "objectivity"—the unmediated access to the natural world that strives to erase all traces of the scientist at work. Art, to the contrary, was based on the cultivation of the imagination and became the unique expression of the artist's inner life. My argument here is that when this break came about, pure mathematics—long the close companion of the natural sciences—sided with the arts. Like them, it abandoned the quest for the correct presentation of reality and became a creative expression of the mathematician's inner life. Imagination, treated with deep suspicion in the sciences, was the essential tool of the mathematician, allowing him access to realms of truth and beauty. The iconography of mathematics, which abandoned the representations of the natural sciences and sided with the arts, was inseparable from this deep conceptual shift. See Daston and Galison, *Objectivity*, 35–38 and throughout.

8. "For me, and I suppose for most mathematicians, there is another reality, which I will call 'mathematical reality.'" See Hardy, *A Mathematician's Apology*, 123.

9. One of the chief advocates of proof by computer is Doron Zeilberger, Board of Governors Professor of Mathematics at Rutgers University and recipient of the Steele Prize, the American Mathematical Society's highest honor. Zeilberger often lists as coauthor of his articles Shalosh B. Ekhad, a computer program he developed that is capable of mathematical proofs.

10. Hardy, *A Mathematician's Apology*, 85.

ACKNOWLEDGMENTS

I wish to thank Margaret Jacob for inviting me to write this book and for her enormously helpful comments on the original manuscript. Peter Galison, Joan Richards, and Spencer Weart also read the original draft, and their suggestions helped make this a far better book. Doron Zeilberger read parts of the manuscript and helped clarify issues in Galois theory, and I am grateful for his help and support. Many thanks as well to my friend Apostolos Doxiadis for inspiring conversations on mathematics and culture, and for helping to put "mathematics and narrative" on the intellectual map. Sincere thanks to my friends and colleagues at UCLA, Ted Porter, Mary Terrall, Sharon Traweek, and Norton Wise, for their continuing support, and for many ideas and suggestions that found their way into this book. Siegfried Zielinski helped shape this book by inviting me to his groundbreaking "Variantology" workshop and giving me the opportunity to interact with a remarkable interdisciplinary group of scholars. Donna Stevens and Jennifer Vaughn of The Planetary Society have been true friends over many years, and this book owes much to their help. Several hours of conversation with Arkady Plotnitsky in Delphi in 2007 opened my eyes to new possibilities of relating mathematics to broader culture, and I am grateful for his insights, as well as his friendship. An intense discussion with Michael Harris on the same occasion helped clarify my own thinking, and I thank him for his perceptive questions and suggestions. Special thanks to all the participants of the "Mathematics and Narrative" conferences and workshops in Mykonos (2005), Delphi (2007), and Budapest (2009), who helped to open up the new field of mathematics and culture to which this book belongs. I am grateful to Ann Downer-Hazell, Elizabeth Knoll, and Matthew Hills of Harvard University Press, and John Donohue of Westchester Book Services, for shepherding this book through all its stages, from acquisition to production.

Daniel Baraz, my oldest friend and intellectual companion, helped shape my thinking, and ultimately this book, in innumerable conversations over three and a half decades. My children, Jordan and Ella, were supportive and patient with their father when he spent hours on his computer, and my thanks, as well as my love, go out to them. Most of all I wish to thank my wife, Bonnie, for her patience, love, companionship, suggestions, critiques, and much much more than I could possibly list here. I dedicate this book to her.

INDEX

Note: Page numbers followed by *f* indicate figures.